HUMAN ASPECTS OF SOFTWARE ENGINEERING

LIMITED WARRANTY AND DISCLAIMER OF LIABILITY

CHARLES RIVER MEDIA, INC. ("CRM") AND/OR ANYONE WHO HAS BEEN INVOLVED IN THE WRITING, CREATION, OR PRODUCTION OF THE ACCOMPANYING CODE IN THE TEXTUAL MATERIAL IN THE BOOK, CANNOT AND DO NOT WARRANT THE PERFORMANCE OR RESULTS THAT MAY BE OBTAINED BY USING THE CONTENTS OF THE BOOK. THE AUTHOR AND PUBLISHER HAVE USED THEIR BEST EFFORTS TO ENSURE THE ACCURACY AND FUNCTIONALITY OF THE TEXTUAL MATERIAL AND PROGRAMS DESCRIBED HEREIN. WE HOWEVER, MAKE NO WARRANTY OF ANY KIND, EXPRESS OR IMPLIED, REGARDING THE PERFORMANCE OF THESE PROGRAMS OR CONTENTS. THE BOOK IS SOLD "AS IS" WITHOUT WARRANTY (EXCEPT FOR DEFECTIVE MATERIALS USED IN MANUFACTURING THE BOOK OR DUE TO FAULTY WORKMANSHIP).

THE AUTHOR, THE PUBLISHER, AND ANYONE INVOLVED IN THE PRODUCTION AND MANUFACTURING OF THIS WORK SHALL NOT BE LIABLE FOR DAMAGES OF ANY KIND ARISING OUT OF THE USE OF (OR THE INABILITY TO USE) THE PROGRAMS, SOURCE CODE, OR TEXTUAL MATERIAL CONTAINED IN THIS PUBLICATION. THIS INCLUDES, BUT IS NOT LIMITED TO, LOSS OF REVENUE OR PROFIT, OR OTHER INCIDENTAL OR CONSEQUENTIAL DAMAGES ARISING OUT OF THE USE OF THE PRODUCT.

THE SOLE REMEDY IN THE EVENT OF A CLAIM OF ANY KIND IS EXPRESSLY LIMITED TO REPLACEMENT OF THE BOOK, AND ONLY AT THE DISCRETION OF CRM.

THE USE OF "IMPLIED WARRANTY" AND CERTAIN "EXCLUSIONS" VARIES FROM STATE TO STATE, AND MAY NOT APPLY TO THE PURCHASER OF THIS PRODUCT.

HUMAN ASPECTS OF SOFTWARE ENGINEERING

JAMES E. TOMAYKO

ORIT HAZZAN

CHARLES RIVER MEDIA, INC.
Hingham, Massachusetts

Copyright 2004 by CHARLES RIVER MEDIA, INC.
All rights reserved.

No part of this publication may be reproduced in any way, stored in a retrieval system of any type, or transmitted by any means or media, electronic or mechanical, including, but not limited to, photocopy, recording, or scanning, without *prior permission in writing* from the publisher.

Acquisitions Editor: James Walsh
Cover Design: The Printed Image

CHARLES RIVER MEDIA, INC.
10 Downer Avenue
Hingham, Massachusetts 02043
781-740-0400
781-740-8816 (FAX)
info@charlesriver.com
www.charlesriver.com

This book is printed on acid-free paper.

James E. Tomayko and Orit Hazzan. *Human Aspects of Software Engineering.*
ISBN: 1-58450-313-0

All brand names and product names mentioned in this book are trademarks or service marks of their respective companies. Any omission or misuse (of any kind) of service marks or trademarks should not be regarded as intent to infringe on the property of others. The publisher recognizes and respects all marks used by companies, manufacturers, and developers as a means to distinguish their products.

Library of Congress Cataloging-in-Publication Data

Tomayko, J. E. (James E.), 1949-
　Human aspects of software engineering / James E. Tomayko and Orit Hazzan.
　　　p. cm.
　ISBN 1-58450-313-0 (hardcover : alk. paper)
　1. Software engineering. 2. Human engineering I. Hazzan, Orit, 1962- II. Title.
　QA76.758.T56 2004
　005.1—dc22
　　　　　　　　　　　　　　2004006047

Printed in the United States of America
04 7 6 5 4 3 2 First Edition

CHARLES RIVER MEDIA titles are available for site license or bulk purchase by institutions, user groups, corporations, etc. For additional information, please contact the Special Sales Department at 781-740-0400.

Orit Hazzan dedicates this book to her parents, Ze'ev (Wolf) and Tikva (Hope) Seger, for the values they inspired in her.

Jim Tomayko dedicates this book to his brother, Jack, for his constant support and for Movie Nights.

Acknowledgments

Jim Tomayko would like to thank Orit Hazzan for commenting on his often stilted prose and doing all the grunt work for this book. Orit Hazzan would like to thank Jim Tomayko for his friendship and the insights and alternative perspectives he so often presents.

We both would like to thank Laura Tomayko for drafting the figures.

We also thank Jim Walsh of Charles River Media for initially contacting us and constantly urging us on with this project; Bryan Davidson for his production work on this book and dealing with our questions; and all the people at CRM who supported the preparation of this book.

Contents

Dedication	v
Acknowledgments	vii
Introduction	xxi

Part I	Software Development Environments	1
1	**The Nature of Software Engineering**	3
	Introduction	3
	Objectives	3
	Study Questions	4
	A Day in the Life of a Software Engineer in a Conventional Company	5
	A Software Engineer's Day at an Agile Development Company	7
	Why Software Projects Fail	9
	Discussion	11
	For Further Review	11
	References and Additional Resources	11
	Endnotes	11
2	**Software Engineering Methods**	13
	Introduction	13
	Objectives	14
	Study Questions	14
	Software Development Methods	14
	Spiral Model	15
	Unified Process	16
	eXtreme Programming	18

Choosing Among the Spiral Model, UP, and XP	22
To Require or Not to Require ... Methods	24
Summary Questions	25
For Further Review	26
References	26
Endnotes	27

3 Working in Teams — 29

Introduction	29
Objectives	30
Study Questions	30
Relevance for Software Engineering	31
Types and Structures of Software Development Teams	31
Democratic Teams	31
Hierarchical Teams	34
Virtual Teams	38
Forming and Rewarding Student Teams	39
Activity	40
Forming Student Teams	44
Evaluation of Student Projects	44
A Game Theory Perspective of Teamwork	45
Outsourcing	48
A Common Use for Virtual Teams: Outsourcing	48
The Capability Maturity Model and Its Use for Outsourcing	49
The Bozo Effect	50
Mind the Gap	51
Summary Questions	52
For Further Review	52
References and Additional Resources	52
Endnotes	53

4 Software as a Product — 55

- Introduction — 55
- Objectives — 56
- Study Questions — 56
- Relevance for Software Engineering — 56
- Software Requirements—Background — 57
- Data Collection Tools — 59
  - Interviews — 59
  - Questionnaires — 63
  - Observation — 65
- Requirements Management — 66
- Characteristics of Tools for Requirements Management — 66
- Discussion — 69
- Summary Questions — 69
- For Further Review — 69
- References — 70
  - Examples of Requirements Management Tools Reviewed — 70

Part II The World of Software Engineering — 71

5 Code of Ethics of Software Engineering — 73

- Introduction — 73
- Objectives — 74
- Study Questions — 74
- Relevance for Software Engineering — 75
- Codes of Ethics — 75
- The Code of Ethics of Software Engineering — 77
- Scanning the Code of Ethics of Software Engineering — 79
  - Activities — 81
- Questions—Principle 1: Public — 83
- Questions—Principle 2: Client and Employer — 84

	Questions—Principle 3: Product	85
	Questions—Principle 4: Judgment	88
	Questions—Principle 5: Management	89
	Questions—Principle 6: Profession	90
	Questions—Principle 7: Colleagues	92
	Questions—Principle 8: Self	93
	Summary Questions	94
	For Further Review	95
	References and Additional Resources	95
	Online Resources	96
	Endnotes	96
6	**International Perspective on Software Engineering**	**97**
	Introduction	97
	Objectives	98
	Study Questions	98
	Relevance for Software Engineering	98
	International Perspectives on Software Engineering	99
	Questions	100
	The High-Tech Industry in Different Countries	101
	India	103
	Israel	104
	Additional International Topics Related to Software Engineering	106
	Women and Minorities in Computer Science and Software Engineering	107
	Summary Questions	111
	For Further Review	111
	References and Additional Resources	111
	Associations that Deal with the Promotion of Women in Computer Science	112

7	Different Perspectives on Software Engineering	113
	Introduction	113
	Objectives	114
	Study Questions	114
	Relevance for Software Engineering	115
	Software Engineering: A Multifaceted Field	115
	The Product versus Process Perspectives of Software Engineering	117
	The Agility Paradigm versus the Heavyweight Approach toward Software Development	119
	Additional Approaches	121
	Summary Questions	122
	For Further Review	122
	References	123

8	The History of Software Engineering	125
	Introduction	125
	Objectives	126
	Study Questions	126
	The Early Days of Computing	126
	Information Hiding—The First Budding of Software Development Methods	129
	Abstraction—Another Part of Methods	129
	The Beginning of Software Development Methods	130
	The Customer's Task in the Early Days of Software Development Methods	133
	Abstraction and Information Hiding Come to the Fore	134
	Software Development Methods Become Part of the Profession of Software Engineering	135
	Objects Arise	138
	Agile Methods Enter the Software Engineering World	138
	For Further Review	139
	References	139

Part III Software-Human Interaction — 141

9 Program Comprehension, Code Inspections, and Refactoring — 143
Introduction — 143
Objectives — 144
Relevance for Software Engineering — 144
Program Comprehension — 145
    Study Questions — 145
    Theories of Program Comprehension — 146
    Activity — 151
Code Inspections (Code Review) — 151
    Study Questions — 151
Refactoring — 154
    Study Questions — 154
Summary Questions — 157
For Further Review — 158
References and Additional Resources — 158
Endnotes — 159

10 Learning Processes in Software Engineering — 161
Introduction — 161
Objectives — 162
Study Questions — 162
Relevance for Software Engineering — 162
Software Engineering as a Reflective Practice — 163
    Conclusion—Software Engineering as a Reflective Practice — 170
    Activities — 170
Learning Organizations — 171
    Activity — 176
Conclusions — 177
Summary Questions — 177

	For Further Review	177
	References and Additional Resources	179
	Endnotes	180
11	**Abstraction and Other Heuristics of Software Development**	**181**
	Introduction	181
	Objectives	182
	Study Questions	182
	Relevance for Software Engineering	183
	Central Heuristics of Software Development	183
	Structured Programming	184
	Successive Refinement	184
	Abstraction	186
	Illumination of Previous Chapters by Abstraction	187
	Additional Topics Related to Abstraction	193
	The Human Aspects of Software Architecture	193
	Architecture versus Design	193
	Quality Attribute Workshops	194
	Metaphors in Science and Philosophy	197
	Abstraction in Computer Science and Software Engineering Education	197
	Summary Questions	198
	References	199
	Endnotes	200
12	**The Characteristics of Software and the Human Aspects of Software Engineering**	**203**
	Introduction	203
	Objectives	204
	Study Questions	204
	Relevance for Software Engineering	205

Software Characteristics	206
Programming Style	207
Abstraction	209
Refactoring	210
Simplicity	210
Evaluation of Programming Style	211
Affective Aspects of Human-Software Interaction	211
Summary Questions	218
For Further Review	218
References and Additional Resources	219
Programming Style Links	220
Endnotes	220
Appendix—Questionnaire	220

Part IV Business Analysis of Software Engineering   223

13 Software Project Estimation and Tracking   225

Introduction	225
Objectives	226
Study Questions	226
Relevance for Software Engineering	226
Poor Software Project Management	226
"Better, Faster, Cheaper"	227
Overtime	228
Avoiding Overtime	229
Historical Data	229
Clark's Method	230
COCOMO II	230
Earned Value	234
The Planning Game	235
Requirements	236
The Method of Up-Front Requirements Elicitation	236

	Requirements Elicitation in Agile Methods	236
	Difficulties for Estimation Caused by Agile Methods of Gathering Requirements	237
	The Team Software Process Development Manager	238
	Playing Games with Estimates and Deadlines	238
	Summary Question	239
	For Further Review	240
	References	240
14	**Software as a Business**	**243**
	Introduction	243
	Objectives	243
	Study Questions	244
	Relevance to Software Engineering	244
	A Brief History of the Software Business	244
	Time-to-Market	247
	Business Cases	247
	Business Plans	248
	Statement of Work	249
	Summary Questions	250
	For Further Review	250
	References and Additional Resources	250
15	**The Internet and the Human Aspects of Software Engineering**	**251**
	Introduction	251
	Objectives	252
	Study Questions	252
	Relevance for Software Engineering	253
	E-Commerce	254
	What Is E-Commerce?	254
	Cognitive Analysis of E-Commerce	256
	Social Perspective of E-Commerce	258

The Timeless Nature of the Internet	259
Communication	261
Summary Questions	263
For Further Review	263
References and Additional Resources	264

Part V Software Engineering Education — 267

16 Case Studies in Software Engineering — 269

Introduction	269
Objectives	270
Study Questions	270
Relevance for Software Engineering	270
Software Management	271
Case 1: Overtime	271
Case 2: Schedule	272
Case 3: Getting New Business	273
Case 4: Discovering Information	274
Software Development Paradigm	275
Case 5: Specifying	275
Case 6: Designing	275
Case 7: Coding	276
Case 8: Testing	276
General Principles	277
Case 9: The Recycling Principle	277
Case 10: Multiple Representations	277
Case 11: Alternative Tasks	278
Case 12: Reflection	278
Case 13: Fingerprints	279
Case 14: Divide and Conquer	279

	Case 15: Finding Hidden Bugs	280
	Case 16: Literacy	280
	For Further Review	281
	References	281
17	**Students' Summary Projects and Presentations**	**283**
	Introduction	283
	Objectives	284
	Study Questions	284
	Relevance for Software Engineering	284
	Case Studies	285
	Construction of Case Studies	286
	Option 1: Construction of a Theoretical Case Study	286
	Option 2: Construction of a Case Study Based on a Field Study	292
	Option 3: Construction of a Case Study Based on an Event in the Past	293
	Presentation of Case Studies	294
	Additional Resources	294
18	**Remarks about Software Engineering Education**	**295**
	Introduction	295
	Objectives	296
	Study Questions	296
	Relevance for Software Engineering	297
	The History of Software Engineering Education	297
	The Education of Software Engineers Today	298
	Teaching Human Aspects of Software Engineering	300
	References and Additional Resources	302
	Endnotes	302

19	Additional Information on Resources Used in This Book	303
	Books	303
	Articles	314
	Web Sites	315
Appendix		**317**
Index		**333**

Introduction

This book focuses on human aspects of software engineering. The rationale for writing this book stems from the fact that the more the software world is developed, the more the software engineering community accepts that the *people* involved in software development processes, not the processes or technology, deserve more attention. In this spirit, this book attempts to highlight the world of software engineering from the perspective of the main actors involved in software development processes: the individual, the team, the customer, and the organization. Indeed, the code and technology are main actors in this process as well, and are discussed in this book. However, when code and technology are addressed, the discussion is conducted from the human perspective.

GOALS OF THIS BOOK

This book is written for software developers, whether university students or practitioners in the software industry. It aims to increase software team members' awareness of the various facets of the human aspects of software engineering. The idea is neither to cover all the available material about the human aspects of software engineering nor to supply a comprehensive and exhaustive list of references about the topic. Rather, our goals in writing this book are

- To illustrate the richness and complexity of the human aspects of software engineering.
- To increase readers' awareness of problems, dilemmas, questions, and conflicts that might be raised with respect to the human aspect of software engineering during the course of software development.

To achieve these goals the book aims to inspire three meta-concepts: *awareness*, *reflection*, and *abstraction*. The idea is to increase software engineers' *awareness* of their development environment, by applying on-going *reflection* with respect to topics whose complexity can be characterized by different levels of *abstraction*. The importance of these three concepts results from the uniqueness of software development environments and processes and the cognitive and social complexities that characterize them.

RATIONALE OF THIS BOOK

If you ask a group of software engineers what software engineering is, you will probably come up with more than one definition, each emphasizing different facets of the discipline. This phenomenon is also reflected in the definitions of software engineering described in professional literature. Indeed, software engineering is a multifaceted discipline. The rationale for this book stems from the fact that software engineering's human aspect, we believe, does not get the attention it deserves.

Software engineering is a young and evolving discipline. The name "software engineering" was popularized by the 1968 NATO Conference in Garmisch, Germany. This conference aimed at discussing the problems associated with the evolving complexity of computing systems, and the acknowledgment of human limitations to cope with this evolved complexity. Indeed, the complexity of software development is acknowledged from the first days of the discipline. For example, [Zelkowitz, Shaw, and Gannon79] explain in the introduction to their book that it is not possible to totally describe in one book all the facets and aspects of software development so that its complexity can be fully understood and appreciated. Interestingly, the nature of the discipline is still questioned today. Sometimes, it is addressed as an art. For example, Knuth, in his book *The Art of Computer Programming* [Knuth97], explains the attractive nature of the process of preparing programs by comparing it to composing poetry or music (this is not surprising, since both music and software engineering have a common core in mathematics). Other perspectives of software engineering emphasize the engineering aspects of the discipline, focusing on software development processes.

The multifaceted nature of software engineering is also expressed by different approaches that aim at describing the *structure* of the discipline. The following are three examples of professional organizations that suggest a mapping of the discipline's body of knowledge:

- The Software Engineering Institute (SEI) at Carnegie-Mellon University (*http://www.sei.cmu.edu/*)
- The SWEBOK (Software Engineering Body Of Knowledge) (*http://www.swebok.org/*)
- The CCSE (Computing Curricula Software Engineering) joint initiative of the IEEE-CS and the ACM (*http://sites.computer.org/ccse/*)

Recently, human aspects of software engineering have gotten more and more attention. Awareness of this component of software engineering appeared in Brooks' book *The Mythical Man-Month* [Brooks75], first published in 1975 and revised in 1995. In the Preface to the 20th Anniversary Edition, Brooks writes that he is surprised that *The Mythical Man-Month* is popular even after 20 years. His statement indicates how difficult it is to apply software development lessons learned to future software development projects. This difficulty may be explained by the multifaceted nature of the discipline and the uniqueness of software development processes. This book addresses this complexity.

We believe that now is the ideal time to discuss the human aspects of software engineering. The high-tech bubble has burst, and it seems that the software engineering community is ready to learn the lessons from that era. It also seems to be the right stage of the evolution of software engineering to consider more closely the people involved in software development processes.

SCOPE OF THE BOOK

As mentioned previously, the importance of the human aspects of software engineering is widely acknowledged today. For example, many failures of software systems are explained by human factors. Taking into the consideration the complexity of the topic, this book focuses on social and cognitive aspects of software engineering, and addresses topics such as teamwork, customer/software-engineer relationships, and learning processes in software development.

It is important to emphasize that this book is not a human-computer-interaction (HCI) book. While HCI focuses on people-software interaction, this book focuses on people-people interaction during the course of software development.

READERSHIP OF THE BOOK

This book is for readers who have some experience in software development. While we cannot quantify this experience, it would be preferable if the reader has already developed a program that is beyond several hundred lines of code and is familiar with software engineering methods. Our working assumption during the process of writing this book is that one needs some experience in software development in order to understand the essence of what is discussed. Thus, for example, many tasks are based on an individual's reflection on personal experience.

STRUCTURE OF THE BOOK

The book consists of five parts. Part I, *Software Development Environments*, addresses the nature of software engineering, software development methods, teamwork, and software as a product. Part II, *The World of Software Engineering*, presents the following viewpoints regarding software engineering: the code of ethics of software engineering, international perspective of software engineering, different perspectives of software engineering, and the history and future of software engineering. Part III, *Software-Human Interaction*, looks at program comprehension, learning processes in software engineering, abstraction and other heuristics of software development, and connections between software characteristics and human aspects of software engineering. Part IV, *Business Analysis of Software Engineering*, examines software project management, software as a business, and e-commerce. Part V, *Software Engineering Education*, analyzes case studies, suggests options for students' summary projects and presentations, and concludes with some discussion remarks about software engineering education.

We now describe each chapter in more detail.

Part I: Software Development Environments

Chapter 1, "The Nature of Software Engineering," deals with the nature of software engineering, emphasizing its human aspects. To illustrate the cognitive and social aspects involved in software development, descriptions of two working days in the life of a software developer are presented to point out how much of such days are about human and not technical aspects. In this sense, these stories establish the rationale of the book.

Chapter 2, "Software Engineering Methods," deals with software development methods. Both heavyweight and agile methods are referred to and analyzed from a human perspective. The aim of this chapter is to illustrate that both technical and human factors should be considered when evaluating what software development method to adopt for a specific software project.

Chapter 3, "Working in Teams," focuses on teamwork, which is a central topic in software development. It presents types and structures of software development teams and discusses issues related to communication within software development teams. A special emphasis is placed on team members' roles in software development teams. In addition, dilemmas that may arise during software development processes and outsourcing related topics are discussed.

Chapter 4, "Software as a Product," brings the software development customer into the picture. In this chapter, customers play a central role and software products are analyzed as merchandise for which one pays. The motivation for this chapter stems from the well-known fact that in many cases, customers do not get the software they require. This reality is partially explained by misunderstandings on the part of the software developers with respect to customers' needs. This chapter suggests tools to overcome such problems. For example, we illustrate how software developers can improve their understanding of software requirements when data is gathered and analyzed by different data collection tools.

Part II: The World of Software Engineering

Chapter 5, "Code of Ethics of Software Engineering," focuses on just that. The essence of the code is presented and different scenarios taken from the world of software development are presented and analyzed. When ethical considerations are intertwined with problems related to human behavior, the right solution is not always simple to figure out.

Chapter 6, "International Perspective on Software Engineering," discusses software engineering from the international perspective. The focus in this chapter is on several events that affected the global information technology market and on two countries, India and Israel, that are unique in some sense with respect to software development. Many software companies have an international market. We believe that a software engineer's awareness of different cultures in general and of different cultures of software development in particular may improve software development processes. The idea is to highlight the fact that software development may be influenced by local and cultural characteristics that software engineers have to take into consideration when they cooperate with people from other countries and other cultures. This chapter also discusses the topic of gender in the information technology sectors.

Chapter 7, "Different Perspectives on Software Engineering," addresses various perspectives on software engineering in general and on software development processes in particular. Our aim in this chapter is to show that there may be different focal issues in software development and to present tools for analyzing different approaches toward the process of software development.

Chapter 8, "The History of Software Engineering," summarizes software engineering from a historical angle and outlines the development of the field from its early days, through the 1968 NATO conference in Germany, until today. This chapter illustrates the dynamic nature of the field. Naturally, human aspects that have influenced its development are emphasized.

Part III: Software-Human Interaction

Chapter 9, "Program Comprehension, Code Inspections, and Refactoring," and Chapter 10, "Learning Processes in Software Engineering," focus again on software team members and their activities during the process of software development. In Chapter 9, the topic of program comprehension is reviewed. It presents different theories of program comprehension and examines code inspection processes. The recent awareness of these two activities reveals that more and more attention is placed on code readability in particular and on the human aspects of software engineering in general.

In Chapter 10, additional learning processes involved in software development are discussed. The rationale for this chapter stems from the continual development of the software engineering world. Consequently, software engineers must continually learn what is new in the field. Specifically, two topics are addressed. The first is software engineering as a reflective practice. Analysis of the field of software engineering and the kind of work that software engineers usually accomplish supports the adoption of the reflective practice perspective to software engineering processes. In this chapter, we examine how a reflective mode of thinking may improve the performance of some of the basic activities of a software engineer. The second part of the chapter deals with learning organizations in general and their implications in the context of software organizations in particular. Knowledge and information are critical assets in software organizations. Thus, they should be managed and thought about in the same manner as many of the more tangible assets.

Chapter 11, "Abstraction and Other Heuristics of Software Development," focuses on software development principles. The discussion is illustrated by focusing on the concept of abstraction. Specifically, connections between abstraction and human aspects of software engineering in general, and the cognitive aspect of software development in particular, are explored.

Chapter 12, "The Characteristics of Software and Human Aspects of Software Engineering," examines the code again. Specifically, connections between characteristics of quality software and the human aspects of software engineering are analyzed. In addition, this chapter examines effective aspects of software development, focusing on debugging.

Part IV: Business Analysis of Software Engineering

The three chapters in this part (Chapters 13, 14, and 15) deal with the business side of software; namely, software project management, software as a business, and the Internet and the human aspects of software engineering, respectively. This view highlights the impact that the business aspect of the software industry may have on the human aspects of software engineering.

Specifically, Chapter 13, "Software Project Estimation and Tracking," discusses main topics, especially time management, of software project management. Chapter 14, "Software as a Business," consists of two main parts: a brief account of how software became profitable and stories of making money with software. The first part of Chapter 15, "The Internet and the Human Aspects of Software Engineering," is about e-commerce. In this chapter, the focus is placed on the end users, closing the circle of developers–clients–end users. Accordingly, while in some of the earlier chapters the developers-clients interrelation has been discussed, in this chapter we explore possible connections between developers and end-users by addressing cognitive and social aspects of e-commerce. The second part of this chapter focuses on the timeless nature of the Internet and its application to the human aspects of software development.

Part V: Software Engineering Education

The next three chapters of the book (Chapters 16, 17, and 18) deal with software engineering education. These chapters wrap up the book with case studies, students' and practitioners' summary projects, and a discussion about software engineering education.

Chapters 16 and 17 summarize the book by case study analysis. Such analysis, we believe, may increase students' and practitioners' awareness, sensitivity, and analytic skills when they participate in software development environments. The analysis is based on theories learned in previous chapters. Chapter 16, "Case Studies of Software Engineering," presents the readers with case studies. Chapter 17, "Students' Summary Projects and Presentations," suggests ideas for students' summary projects and presentations. Each student (or a group of students) is asked to select a topic that belongs to the human aspect of software engineering and to identify an interesting problem to be researched. They are asked to research the topic, to analyze data based on the theories discussed in this book, and, based on that exploration, to construct a case study. Chapter 18, "Remarks about Software Engineering Education," comments on software engineering education in general and about the organization of a course that deals with human aspects of software engineering in particular. Chapter 19, "Additional Information on Resources Used in This Book," closes the book.

STRUCTURE OF EACH CHAPTER

Each chapter is presented in a way that enables instructors to use it for reading material, class activities, questions for class discussion, and homework assignments which make it easier for practitioners to gain insights. All chapters have a similar structure. Each is composed of three parts: an introduction, the body, and a summary.

Each chapter starts with an *Introduction* that presents in brief what is going to be discussed in the chapter. Then, the chapter's *Objectives* are outlined. The list of objectives reflects what learners may gain in the process of and after learning the presented topic. Just after the *Objectives,* several *Study Questions* are presented. The aim of these questions is twofold: first, to raise students' and practitioners' interest in the topic; and second, to invite learners to start thinking about the topic before it is presented to them. The *Study Questions* are not restricted to technical questions and sometimes invite reflection processes and case study analyses. The introductory part of most chapters ends by discussing the *Relevance* of the discussed topic to *software engineering*. This section explains why the discussed topic is included in the book and highlights the topic's relevance for the daily life of software developers.

The body of each chapter presents the material in a way that puts the developers at the center of the discussion. The discussion examines what happens to the developers, how they may behave, how they may react, what feelings they may have, what problems they may face, how they may solve the problems, and so forth. During the presentation of the material, questions and reflections are intertwined. The idea is to ask readers to stop reading from time to time to think about what they have already read, reflect on it, and put what is learned in a wider context. These tasks aim at equipping readers with problem-solving tools to be used when they face new problems in software engineering. Accordingly, based on the theoretical material presented, many questions invite the reader to analyze situations in software development, to create scenarios that may raise problems, and to suggest solutions to these problems.

The summary part of each chapter integrates all that was learned in the chapter. In most chapters, the *Summary Questions* ask the reader to examine the discussed topic from a global perspective, viewing it as a comprehensive subject. The *For Further Review* questions aim at raising topics in which different aspects of what is discussed in the chapter can be used. These questions increase students' and practitioners' involvement, thinking, and understanding of the topic. These assignments invite the readers to apply what is presented in the chapter and usually require deeper thinking and analysis. The summary part of each chapter ends with a list of *References* and, when appropriate, *Additional Resources* for those readers who want to deepen their knowledge.

HOW TO USE THIS BOOK

This book can be used as a course textbook and for practitioners.

As a textbook, the book can be used in software engineering departments for courses that deal with software engineering methods in general or with human aspects of software engineering in particular. There are no specific prerequisites are required. However, as mentioned previously, it is recommended that learners who use the book have some experience with software development. Such experience may improve their understanding of the discussed topics.

The different chapters of the book can also be used separately in other software engineering courses when instructors want to highlight human aspects of a specific topic. For example, Chapter 8, which presents the history of software engineering, may be integrated naturally into a Methods of Software Engineering course for highlighting the course material from the historical perspective.

In industry, the book can be used for software engineering practitioners. In this case, practitioners can use it when they want to learn a specific topic or increase the awareness of their software teams to human aspects of software engineering. The second goal can be achieved, for example, by organizing a seminar about the human aspects of software engineering based on selected topics presented in the book that are relevant to the particular software development environment.

Whether the book is used in academia or in industry, the order in which the chapters are presented is only a suggestion. Because the various chapters are self-contained yet sometimes dependent, they can be learned in almost any order. To help readers navigate, relationships between different chapters are mentioned when appropriate.

REFERENCES

[Brooks75] Brooks, P. Fredrick, *The Mythical Man-Month: Essays on Software Engineering*, Addison-Wesley, 1975.

[Knuth97] Knuth, Donald E., *The Art of Computer Programming*, Third Edition, Addison-Wesley, 1997.

[Zelkowitz, Shaw, and Gannon79] Zelkowitz, Marvin V., Shaw, Alan C., and Gannon, John D., *Principles of Software Engineering and Design*, Prentice Hall, 1979.

Part I: Software Development Environments

Chapter 1, The Nature of Software Engineering
Chapter 2, Software Engineering Methods
Chapter 3, Working in Teams
Chapter 4, Software as a Product

1 The Nature of Software Engineering

In This Chapter

- Introduction
- Objectives
- Study Questions
- A Day in the Life of a Software Engineer in a Conventional Company
- A Software Engineer's Day at an Agile Development Company
- Why Software Projects Fail

INTRODUCTION

Software development and engineering is unlike many professions. Even though it is a technical profession, people issues surrounding both development and management are equally, if not more, important. This book is primarily about the human side to software engineering, and this chapter sets the stage.

OBJECTIVES

- Readers will understand that most of the reasons for software development success and failure are people-, not technology-centered.

- Readers will be able to discuss the effects of human interaction in software development processes.

STUDY QUESTIONS

1. How did *software engineering* become a term?
2. Is there a good technical solution to software development problems?
3. How and why are agile methods considered more people-affirming?
4. Compare software engineering with other professions with which you are familiar. Reflect: What aspects of the professions did you compare?

"Software engineering is boring!" Guess who said this? Is it a senior nearing graduation? Is it a disgruntled undergraduate? No! It is a colleague of one of the authors, a Turing Award winner, one of the most intelligent people in our field. He cuts right into the heart of the subject of this book: if we concentrate only on technical aspects of software engineering, it truly is boring; if we pay attention to the panoply of human activities, it becomes much less so.

Software engineering is a lot like civil engineering, in that many practitioners believe in finding a couple of fundamental tools to solve any problem. Civil engineers can trace most of their solutions to the fields of Statics and Strength of Materials. Chemical engineers concentrate on ever-improving processes, and aeronautical engineers are just now getting powerful predictive tools, after nearly a century of trial-and-error flying.

Software engineering tried to find a "best solution" for much of its early existence. The North Atlantic Treaty Organization (NATO) conference at Garmish, Germany, in 1968, at which the term *software engineering* became popularized[1] was most remarkable in that the attendees came together to share solutions, and wound up sharing problems.

Then followed 20 years of searching for "one best way" of doing things. This seemed to end with Fred Brooks' "No Silver Bullet" essay in *IEEE Computer* magazine in 1987 [Brooks 87]. In this article, he pointed out that there appeared to be more than one solution to any software-engineering problem. His reputation was so great that he put the skids on the Computer Aided Software Engineering (CASE) solutions then in vogue.

His article raised the old "management versus technology" debate in which some hold that software failures (still in the majority of projects in the industry) were caused by management (working with people) issues, rather than by technology. Brooks himself, at a Software Engineering Institute (SEI) curricular workshop, said that things done by large groups—formally—were not as successful as those done by small groups—informally. One of the best examples is PL/I versus Pascal. The former is a combination of COBOL and FORTRAN, two languages built by committees giving structure to a third. Niklaus Wirth built the latter. PL/I never took off, either as a teaching or commercial language. Pascal did as both. More milestones of the software engineering history are presented in Chapter 8, "The History of Software Engineering."

At this stage, perhaps one good way to compare more traditional software development is to contrast the way two engineers spend the day. The first is at a conventional workplace, and the second at a workplace with less formality.

TASKS

1. What is the "management versus technology" debate?
2. Who is Fred Brooks?
3. What is a "silver bullet?"
4. What is CASE? What does it do?

A DAY IN THE LIFE OF A SOFTWARE ENGINEER IN A CONVENTIONAL COMPANY

Joe walked the long hall past the giant aluminum shaping machines, flinching from the noise. Despite the discomfort, he was happy to see the result of his work pile up at the end of the assembly line. Veering to the right halfway down the big building, he came to the door of the "East Lean-to," as it was called. Entering a code into numbered buttons near the knob, he entered. What appeared to him then were two sets of cubicles separated by a wider aisle than was usually between rows. The right set was nearly empty. It had housed as many as 50 engineers, working on a project some thought upper management used as a loss leader. Now only a few testers were left, the others having been transferred to other projects as this one wound down. The project was already a year late.

The other cubicle group was mostly filled by colleagues on his project. There was one secretary for his team, sitting in an open space in front of the cubicles. They smiled to each other and he made his way to his cubicle amid a storm of greetings from others in his row.

Joe threw down a few papers and tried to clean his cup by running a finger around the inside. He then filled the cup from a never-ending urn of bitter coffee. Making his way back to his cubicle, he looked inside. There were two pieces of furniture available: a desk, the surface of which was halfway occluded by a monitor, keyboard, and mouse; and a shelf on one wall. Once, before time-sharing, the surface space was enough, because printouts could be spread out on the desk. Now, the computer hardware was too big (the company was at least a decade from making flat panel displays ubiquitous) to fit on the shelf, so a lot of real estate was used on the desk.

Joe sat down, tried to figure out where he stopped the previous day, and began coding his task for a couple of hours. Promptly at 10 A.M., whether or not they were at a natural stopping place, his row left their cubicles and made their way to the cafeteria. The plant held too many workers for the number of seats in the cafeteria, so the room was time-shared, each area of the plant having strict break and lunch times.

Joe made his way back to his cubicle with a sugar buzz—not too unpleasant. After scoring a diet soda from a dispensing machine, he found one of the senior engineers in his cubicle. The senior engineer wanted a verbal introduction to the automated test suite. About an hour later, they ended the conversation.

Joe took out some code that was to be reviewed after lunch. He clocked his time, thinking that it was good he had to report "how long" and not "when." About an hour later, a friend from a different project with the same lunch shift came by and they found their way to the cafeteria again for the half-hour break.

Right after lunch, the code review (see Chapter 9, "Program Comprehension, Code Inspections, and Refactoring") was scheduled in the windowless conference room adjacent to his cubicle area. Four persons were present: a moderator, to keep attendees from exchanging blows and keep the meeting moving along; a reader, to paraphrase aloud or read the code verbatim; the developer, who explained obscure sections of the code and recorded defects; and Joe, who, like the others, looked for defects. After all told the moderator how much time they spent preparing for the inspection, the reader began by paraphrasing the first 10 lines of code, and they spent nearly the next two hours working through the logic on their segment, sometimes spiritedly.

When Joe got back to his desk, it was nearly 3:00 P.M., time for his daily inventory of e-mail. There were only 72 messages today. He was one of the few who checked e-mail as rarely as daily, or sometimes twice daily, the minimum for others. He also turned off the "beep" when a new message came in. First, he went down the title and return address list looking for obvious spam. He quickly deleted irrelevant mail, pausing to click the REMOVE link if one was present. Then he started down the remaining list of about 50 or so. About half concerned his project, and the others were of general interest or personal contacts. An hour later he had cleared the message queue, something that was rarely done. He figured that if he saw mail only once or twice a day, he would clear his queue and keep nobody waiting too long. Still, he got the occasional mail asking if he had received a message sent two hours earlier.

The last hour and a half of the day he spent going over the code he wrote in the first two hours. He looked again for places to insert comments, and then brought all the other concerned documents up to date, finishing a few minutes past quitting time. He left the cubicle area for the walk to his car, again passing the raucous assembly line, with the second shift now on duty. The "Sweathog Stampede," as some called it, of first-shift machinists had already left. The departure of the engineers was timed to avoid the big exchange. As Joe unlocked his car, he had already filed work away in the back of his mind.

TASKS

1. List all the human contacts Joe had during his day. How much time did he spend on them versus working alone on his project?
2. What proportion of the human contacts had to do with the project?
3. Would you like to work at the same company as Joe? Why or why not?

A SOFTWARE ENGINEER'S DAY AT AN AGILE DEVELOPMENT COMPANY

Jeanne parked her car in the small lot next to the modern rectangle of her company's utilitarian building. She trotted toward it, past the security guard at his desk, and into the open workspace her team occupied. The room was nearly one-fourth the area of the first floor. There were cubicles across the end opposite the entrance. Between her and the cubes were several desks with nothing under them to bang knees and shins against. On top of these desks were one keyboard, monitor, and mouse, with two chairs in front of them. Windows covered one entire wall, giving an airy, well-lit feeling. On the windowless wall was a larger white board; the left half contained a drawing of the software architecture of the product, at least as it was two days ago. The right half was newly cleaned.

Going to the area near the white board, she glanced at the Integration Machine, on which resided the code done so far. She went to the other end of the table on which it sat, where there was small refrigerator. Jeanne took a diet soda from it and several pieces of chocolate from a bowl on top of the refrigerator. There was also an open bag of pretzels from a meeting yesterday, but she ignored it.

Taking her drink and the chocolates, she went into her cubicle. There she zipped through the morning's e-mail. None of it was about her project. She also took advantage of the telephone in the cubicle to call her doctor's appointment line, which had just opened.

Leaving the cubicle a few minutes later, she went to the area where the daily standup meeting took place. The rest of her development team had arrived by then. It was called a "standup" meeting because everyone stood for its duration. The team went over the tasks it accomplished yesterday, and someone distributed to task owners their final task cards of this iteration.

Jeanne would have to wait to do hers in the afternoon, as it was not her turn to pick a partner. She noticed that another engineer indicated that they should pair.

The rest of the team also paired off. It was like some of the media-depicted drug-induced pairings at 1960s parties: anyone would do, and you could not say "no."

The other engineer was roughly equal to her ability. She was happy that was true today, as she did not want to drag along a junior engineer, or be dragged along by an old hand. They picked a desk and went to work. As he settled in front of the monitor, keyboard and mouse at hand, she got another can of caffeine. With a glance at the task card, she sat in the other chair, moving it to the other engineer's left.

They worked for the rest of the morning on their task. She pointed out five logic errors and about ten typos. When she felt like it, the drinks in the refrigerator and the food atop it, augmented by two new bags of snacks, were hers for the taking. About midway through the morning, her partner took a break to go to his car to retrieve something personal he had forgotten.

They started the morning together by writing several tests for the functionality of their unit. Just before noon, the tests all showed a green passing bar when run. Her partner took a diskette with their work to the Integration Machine. All the tests from previous coding sessions ran the first time. Their code was officially integrated a few minutes after noon.

Lunch was next on the day's agenda. They went to an inexpensive family restaurant in a nearby strip mall. Most of the team was at a rear table. The pair joined them, and what ensued was raucous laughter interspersed with eating.

Returning to the workroom, Jeanne glanced again at her task for the day and the architecture on the wall. She chose a team member from the available pool and prepared to pair. She took her seat in front of the monitor and keyboard. Her partner had been with the organization a bit longer and so should have more experience. Jeanne liked to pair with more senior people when she was "driving." She felt that it was better to pair with someone from whom she could learn. Right now, her partner slouched to her left, sipping on a box of juice and looking a lot younger.

He watched Jeanne's tests take shape, making a suggestion or two. Then she methodically wrote code to pass all the tests, finishing just before quitting time. She tested her new code and was able to integrate it. There was a short line at the Integration Machine, since it was the end of iteration.

When all the new code was integrated, a cheer went up. The iteration was finished. The coach, the most senior employee there, led the team to a bar in the strip mall and, for an hour or so, the team celebrated the end of their iteration. Then, those who drank had an extra hour or so to clean up the code. Those who did not (like Jeanne, who nursed yet another diet drink), went home.

The Planning Game for the next iteration (see Chapter 2, "Software Engineering Methods") was in the morning. Tomorrow was soon enough to think about it.

TASKS

1. List all the human contacts Jeanne had during her day. How much time did she spend on them versus working on her project?
2. What proportion of the human contacts had to do with the project?
3. Would you like to work at the same company as Jeanne? Why or why not?

WHY SOFTWARE PROJECTS FAIL

We described two widely differing development days. The first is still the most frequently found; the second contains several ideas due to be tried out by 60 percent of the Information Technology (IT) shops in the immediate future. Frequently, this is why projects fail because required documentation hides how the project is doing.

We can see in Figure 1.1 that the project slowed when it reached "90 percent done." We call this "90-percent dumb," because there is no way to tell if it is true. Does it mean 90 percent of the documentation, the code, or the completed solution? Note that the project seems to be steadily marching toward completion. That is, until the 90-percent barrier is reached. You can almost hear the project manager: "We're 90 percent done." "We just have a few bugs to wring out." "We do not know exactly how long removing the defects will take, but we're over 90 percent done." "We're 98 percent done." And in the final week: "We're done." Nearly half of the project's elapsed time is often spent being 90 percent done.

Software development is nonlinear. As an example, a consultant friend of ours is out of town a lot. A new fence project at his home was postponed for over a year. He got a neighborhood boy to dig the postholes. Late in the afternoon, having dug all but one hole, the boy came to the door and said that he needed to quit and clean up, could he have the pro-rated pay for the holes he dug? The boy said that he would come back the next Saturday and finish. The consultant did as he was asked, but the boy never again showed up. A few weeks later, the consultant had some free time, so he went out to dig the last posthole. He found a boulder that would take more time to extract than the time it took to dig the other postholes. That is what we mean by "nonlinear." People rarely have any idea how far along they are with this way of measuring development. They can be 100 percent done with 90 percent of the tasks, and the next task could take 50 percent of the project time.

Looking again at Figure 1.1, earned value tracking can take care of this uncertainty. Using earned value, your project does not get credit for a task until it is completely done. In other words, if the task is to document the architecture, using earned value in agile methods you would not get credit for documentation until the last iteration. Appearing constantly on a white board counts for nothing. Therefore, every

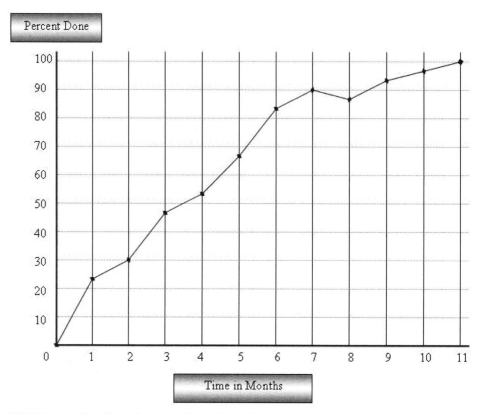

FIGURE 1.1 The life cycle curve of a typical software development.

iteration may be done 100 percent, but the tasks in an iteration may be worth only 20 percent of the tasks on a project. For example, doing the postholes the first Saturday resulted in the boy doing 100 percent of his work that day, but only about half of the total task. When somebody finished the last hole, that task was 100 percent complete, but a few weeks late. By having concrete tasks and subtasks, actual progress can be measured, and "90 percent done" can really mean something. Some years ago, Dick Tausworthe, of NASA's increases, the accuracy of the estimate increases as well [Tausworthe81].

One cause for software failure, then, is blown schedules for chunks of the project, making those parts of the project almost impossible to finish on time because their estimates of completion are so poor. Nonlinearity takes its toll. This book contains stories of software successes and failures. Looking at them, we find that small granularity of tasks and frequent iterations are the key to success. The next chapter talks about some software development methods that have these characteristics.

Obviously, additional reasons and factors influence the success of software projects, such as customer involvement in the development process.

TASKS

1. How does earned value work?
2. What is Tausworthe's Principle?

Discussion

What physical changes have to be made in the workplaces in the case studies?

FOR FURTHER REVIEW

Based on the two case studies presented in this chapter, construct principles of an ideal place to work (for software development).

REFERENCES AND ADDITIONAL RESOURCES

The case studies in this chapter are from the authors' experiences and that of Francesco Cirillo, of XP Labs, Inc., in Italy.

[Brooks87] Brooks, P. Fredrick, "No silver bullet: Essence and accidents of software engineering," *IEEE Computer* (April 1987): pp. 10–19.

[Tausworthe81] Tausworthe, R. C., "Deep Space Network Software Cost Estimation Model," *Publication 81-7*, Jet Propulsion Library, Pasadena, CA, 1981.

Earned Value Web site: *http://www.acq.osd.mil/pm/*

ENDNOTES

[1] There is some evidence that Douglas Ross used the term "software engineering" in a course at Massachusetts Institute of Technology (MIT) before the NATO conference.

2 Software Engineering Methods

In This Chapter

- Introduction
- Objectives
- Study Questions
- Software Development Methods
- Choosing Among the Spiral Model, UP, and XP
- To Require or Not to Require . . . Methods
- Summary Questions

INTRODUCTION

The products of software engineering are software systems. As with other engineering areas, products should be developed following some methods. This chapter is about methods that can be applied for the process of software development. As described in Chapter 8, "The History of Software Engineering," most of the methods used for software development are based on the activities of specifying, designing, coding, and testing. However, these activities are implemented differently by different software development methods.

This chapter focuses on the implementation of these activities by three methods of software development: Spiral Model, Unified Process, and eXtreme Programming. With respect to each method, we outline its main idea and rationale, the main activities on which it is based, and its analysis from the perspective of the human aspect. Readers who choose to learn now about the evolution of these methods are welcome to read Chapter 8 first and then return to this chapter.

OBJECTIVES

- Readers will understand the shape of the model involved in each method.
- Readers will be aware of human aspects of each method.

STUDY QUESTIONS

1. What are the differences among the Spiral Model, Unified Process, and eXtreme Programming that would influence your choice of one over the other for development?
2. How many of the reasons in the answer to Question 1 are based on the methods' human characteristics?

SOFTWARE DEVELOPMENT METHODS

It is well known that software development is a complex process. This process suffers from problems such as budget overruns, late delivery, and, maybe most important, failing to meet the customer's needs.

In his famous article, "No Silver Bullet" [Brooks87], Frederick Brooks presents his argument that even though the software development community seeks "a silver bullet"—a software method that may solve such problems—such a bullet is not foreseen. However, because software should go on being developed (whether we have the silver bullet or not), this chapter focuses on software development methods highlighting their human perspective.

Specifically, we focus on three software development methods: Spiral Model, Unified Process (UP), and eXtreme Programming (XP). There are other software development methods, but these three are accepted today by many software practitioners and are used by many software houses. Our intention in this chapter is neither to teach the development methods nor to outline their details. There are many software engineering books that address software development methods. Rather, our intention is to illustrate the human aspects of these development methods and to increase the reader's awareness of these aspects when they come to evaluate and adopt a specific development method. Because XP emphasizes the human aspect of software development, the majority of this chapter discusses this method.

For each method we describe how the four basic activities of the paradigm of software engineering—specifying, designing, coding, and testing—are implemented. The rationale for this set of activities, on which most of the accepted software development methods are based, is explained in Chapter 8, where the history of software engineering is presented. In addition to having common characteristics, the three methods implement abstraction and information hiding (see Chapter 8), and all use iterations; that is, the development process is incremental.

We chose this organization to reflect the fact that though a set of activities are conceived and accepted as the core of software engineering, they may be implemented in different ways (depending on people's orientation and preferences, types of projects, etc.). We hope that this organization will help those who are interested in analyzing other methods of software development.

Spiral Model

The Spiral Model of software development is defined by Barry Boehm [Boehm88]. Primarily, Boehm thought of his method as a risk-reducing method. It is called the Spiral Model because the software continually gets larger as each increment goes through the activities of specifying, designing, coding, and testing several times.

The Spiral Model begins with engineers trying to build the least-known kernel of the software. The idea is that the part that needs most exploration is the riskiest. A portion of the risk is poor time estimation because we do not know what it will take to build the software. Another portion of the risk is the sheer unknowns. The first spiral through software development is meant to respond to these risks, and, if possible, to build an active kernel on which later software can hang (see Figure 2.1). Engineers should then decide which is the next most risky component and build it, going through the activities again. This cycle is repeated until the software is complete.

Risks are addressed in decreasing complexity. The software is "grown" rather than just developed. The Spiral Model is the first complete capturing of the principal characteristics of modern software engineering. It is iterative and goes through specifying, designing, coding, and testing during each iteration.

The Spiral Model is one of the last modifications to the Waterfall Model (see Chapter 8). As such, it has the nascent origins of human aspects. Most are encapsulated in risk reduction. Humans worry a lot about risk. The Spiral Model, in reducing risks, eases the minds of the managers who choose it.

16 Human Aspects of Software Engineering

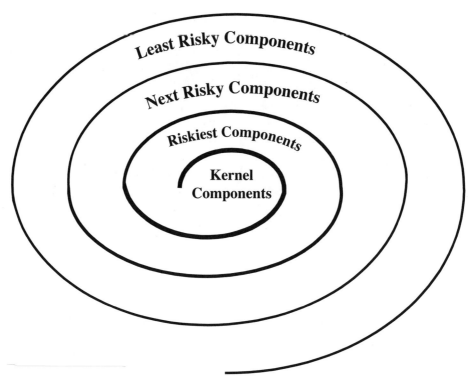

FIGURE 2.1 The Spiral Model.

TASKS

1. Discuss how the Spiral Model reduces risk.
2. What does risk mean in the project you work on (whether it is developed in the university or the industry)? What risks do you face? How do you reduce these risks?

Unified Process

The Unified Process (UP) bills itself as use-case driven, architecture-centric, iterative, and incremental. The third characteristic, and consequently the last, are a result of the software development paradigm (see Chapter 8). Use cases and architecture pervade the UP.

Use cases are *what* the user wants a computer system to do, not *how* to do it. Formal descriptions of use cases contain pre-conditions and post-conditions. Actors in use cases can be either persons or other systems. Both are depicted in the

Unified Modeling Language (UML) as a stick figure. As an example, a withdrawal at an automated teller machine (ATM) could be one type of use case; a program that needs reservations from a flight is another. In a way, use cases represent the object-oriented view of the world. Objects represent real-world "things" just as use cases reveal real-world activities and relations.

UP takes advantage of these things. It is often misnamed the "Rational Unified Process" (RUP) since Rational has done more to popularize it. Rational was bought by International Business Machines (IBM) in February 2003.

UP is the international standard for development, especially software development. It consists of running the paradigm activities of software development through several iterations, one or more of which are phases (see Figure 2.2).

Eighty percent of the use cases are made during the Inception and Elaboration phases, which drive part of the shape of design with minimal code and tests. The use cases represent the highest risks, and the requirements can be derived from them. An important product at the Initiation phase is a Vision document for the software.

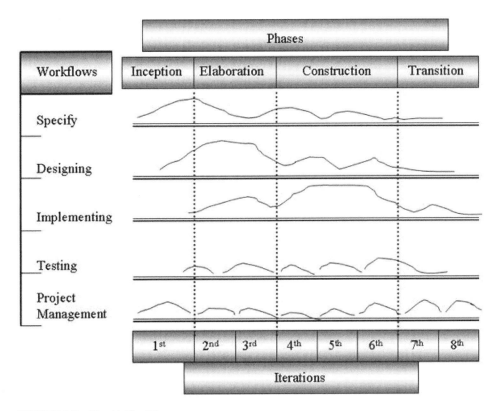

FIGURE 2.2 The Unified Process.

At the Elaboration Phase, more of the use cases are done, engineers revise the architecture, if needed, and more code and tests are completed. The Use Case View and some of the Analysis View of the Unified Modeling Language (UML) help here.

TASK

Select a system with which you work on a daily basis. Identify what use cases define this system.

The next phase, by now probably the third iteration, is the Construction Phase. The project ramps up in personnel to actually build the software. Although Brooks' Law [Brooks95] states that adding personnel to a *late* software project will make it *later*, adding *planned* expansion to an *on-time* project is all right.

The last phase is the Transition Phase, which contains activities often done in other projects, but not part of most methods. These activities include training, begun in earlier phases, and installation. Few methods actually spell these out like UP does.

Basically, the architecture is driven as a complement to the use cases. Some use cases represent business drivers, and thus are key contributors to the architecture. Requirements and the Vision statement are products of use cases. The architecture is more refined during the Elaboration Phase, and little ought to be done to it in either the Construction or Transition Phases. Specifying, designing, coding, and testing are done as part of the "workflows" of UP, each assigned to a person with a specific job. They know exactly what to do in each phase [Jacobson99].

The human aspect of UP begins with the use cases. Each shows something that will be done to the user when he interacts with a component of the software. Since the architecture is extracted from these interactions, there is plenty of opportunity to make the software relevant to humans.

TASK

Discuss how going through the four phases of UP (inception, elaboration, construction, and transition) enlarges the roles of the four activities of software development.

eXtreme Programming

eXtreme Programming (XP) is one of the agile methods of software development [Beck00]. In general, agile methods are more tied to the human aspects of development (see the *Manifesto for Agile Software Development* at *http://agilemanifesto.org/*). XP is based on four values that are expressed by 12 practices. The four values are communication, simplicity, feedback, and courage. The 12 practices are the planning game, pair programming, refactoring, simple design, continuous integration, test-first programming, collective ownership, coding standards, short releases, a metaphor, sustainable race, and an on-site customer. Many times, the practices are shown as dependent on each other [Beck00, p. 70].

The reason XP is called "extreme" is that the volume of each practice is "turned up." Let's not just have periodic access to the clients; let's offer to let the clients move into our workplace and communicate their goals, even though that might mean that requirements change. Let's not just program with occasional inspections; let's pair off, using one keyboard and mouse for each pair, and open ourselves to continually finding defects [Williams and Kessler02]. As can be seen, "turning up the volume" is a metaphor for taking practices to their extreme.

A key concept of the XP method is that of bringing a more humane experience back to programming. With other methods, people who just want to program start by looking at a mountain of documentation and analysis and putting code on paper. XP, as befits a method designed for change, emphasizes rapid delivery of business value to the customer. Therefore, some code is inevitably given to the customer very soon.

There really is no project manager in XP. The closest anyone comes is the Coach. The Coach is usually some senior person, familiar with XP, who keeps everyone on process. Collective ownership of the code, to the point that someone could change someone else's code, works against having hierarchical organizations (see Chapter 3, "Working in Teams").

Pair programming is just that. Each pair codes at one workstation: the "driver" works the keyboard and mouse, the "navigator," looks for defects. Communication can take place at a micro-scale between these two. Traditional managers are often horrified, feeling that pair programming takes twice as much time and effort. In fact, however, it takes less time to fix any defect [Tomayko02], and there is increased corporate knowledge about the product. This leads to knowledge sharing, and if one person leaves the team, the development can go on smoothly, as no private knowledge has been retained.

TASK

Look at the pair programming Web site (*http://www.pairprogramming.com/*). Summarize the main findings about pair programming presented there. What are your conclusions?

The client and the developers, fostering macro-level communication, play the Planning Game. The customer comes to the table with a deck of cards, one high-level "story" per card. The deck is given to the team, which spreads out the cards among themselves, choosing what they are going to do. Their move is to then figure out what tasks make up each story and estimate its length. They then give the cards back to the client, who indicates which stories are most important from the business value standpoint. The team's second move is to give the prioritized stories back to the developers. The developers, knowing their "velocity,"[1] figure whether they can do the indicated stories in the time of the planned iteration (two to three weeks). They then pass back the lowest priority stories for the client to confirm that is the case and that the customer will receive software that has the highest current value. Then, the developers do load balancing among individuals to further ensure that they can finish the selected stories in the planned iteration. Note that the responsibility for task development is personal, although the development is done in pairs. Furthermore, it is important to note that at the end of each planning game, each developer knows that each team member has the same load. This is very encouraging. This feeling, together with the fact that the responsibility for the development task is personal (although the development itself is done in pairs), almost ensures that free riders do not enter XP teams.

TASK

Discuss connections among "load balance" and team work (see Chapter 3).

Empowered, the developers then go to work. There is more contact with the actual people who do the work than clients often have. Similarly, there is more contact with the customer than developers usually have. Note that the developers wind up working on exactly what the client wants now. In addition, as the developers take part in the planning game, they all know the role of each feature they develop (and should not derive this information from a written document).

The Human Aspects of eXtreme Programming

Of the 12 practices of XP, many are humane, particularly the "40-Hour Week" (or sustainable pace, or no extended overtime). The practice is based on the assumption that tired programmers make more mistakes.

However, it is in the four values of XP that we find the human aspects of XP most clearly. The first of these is communication. That value is encapsulated in pair programming. Moreover, communication is accomplished by the metaphor, simple design, on-site customer, the Planning Game, and the coding standard.

Members of pairs can speak with each other, and do. They wind up being the "grapevine" for the project, as change is reflected immediately and constantly in their work. They are the ones who bring the architecture up to date, or modify the database schema.

The metaphor is developed early. It is meant to provide a common language for both developers and clients to discuss the software. As it becomes more entrenched into the project, it leads both developers and clients toward architecture to guide the work of the pairs.

"Simple Design" will show up again under a different value. Here, it means the design of software architecture, and is easy for the pairs to understand. This way, the pairs can tell at a glance if they are doing work that leads to eventual success.

The Planning Game will also show up later, but here it is a way to communicate individual velocities and customer priorities.

The on-site customer is even more of a communication tool than the pairs. Do you need something clarified? Just shout across the room. This naturally leads to the production of software that the customer needs, not software that the developers assume the customer needs. Indeed, more than once you will hear developers claim that customers do not know what they want. XP helps developers reveal these needs together with the customer in a gradual process.

A coding standard is another aspect of communication. The theory is that everyone looking at the code will see the same style. It is not important what the coding standard looks like, more that there is one. For example, Java has a coding standard on the Sun Web site. Use it! It is okay as long as pairs can move around the code and not be confused as to style. It is clear how naturally the practice of coding standards supports the practice of collective ownership.

Note that communication subsumes most XP practices. Indeed, communication is very much a human characteristic of software engineering.

Simplicity is the second XP value. It is most often exemplified in, of course, simple design, and refactoring (see Chapter 9, "Program Comprehension, Code Inspections, and Refactoring"). The main purpose of refactoring is to make the code simpler. It turns out that "simpler" is hard to do. Constant rewriting of the code, which is refactoring, and simplifying the design as the project progresses are a part of simplification. Another important part is always doing the simplest thing that results in value returned to the customer. Sometimes, people skip refactoring, because it does not add functionality to the code. However, the added value that refactoring has is expressed in later stages of the development when the code is better understood, more readable, and more manageable.

Feedback, the core of modern systems theory [Mindell02], is the third XP value, shown by test-first programming, short releases (frequent builds), continuous integration, and the on-site customer. Test-first programming is the most direct and shortest form of feedback. It shows how an individual unit can be made better and whether the thing that we are working on is functioning. Short releases tell us whether the system is progressing. The customer is there to keep us honest. This value is the closest to at least one tenet of "reflective practice" in XP (see Chapter 10, "Learning Processes in Software Engineering").

The final value is courage, which is sometimes difficult to describe. It means that optimism is well placed. The 40-hour week (sustainable pace) falls under courage, as does the Planning Game. Basically, we believe that everything will turn out all right in the end. We believe we will find any requirements changes from our customer, and will be able to implement them; hence, no overtime. Everything we decide to do in the Planning Game will be done, if not in this iteration, then in the next. Courage is also expressed by other practices. Courage is needed for refactoring code that has already proved to run and pass all tests; when the ownership of the code is collective, each pair needs the courage to modify code that another pair wrote. Because all these activities are supported by the atmosphere that XP inspires, it is obvious that they are conducted by all team members in order to improve the developed software.

As is illustrated previously, some of the XP practices outline specifically how to behave (e.g., pair programming, sustainable race), others specify what to do (Planning Game, refactoring, continuous integration, test-first programming, coding standards, on-site customer), and yet others may be conceived of as inspiration practices (simple design, collective ownership, short releases, a metaphor). The practices in the first group tell us how to interact with others and with ourselves; the practices in the second group outline very specific guidelines how to carry out the development process; and the practices in the third group are guiding principles that inspire the actual development.

TASKS

1. What are the XP practices that contribute to communication within teams?
2. What are the three most human aspects of XP, and why?
3. Identify connections between the 12 XP practices; that is, how might the performance of one practice influence the performance of other practices?
4. In this section about XP, we present two organizations of the XP practices (according to the XP values they support and the kind of instructions they give). Suggest another organization of the XP practices.

CHOOSING AMONG THE SPIRAL MODEL, UP, AND XP

There are different criteria for choosing a software development method. From the human perspective, the main criterion of choice is communication. There is little doubt that persons using XP communicate better within a team. Therefore, the Spiral Model and UP can be used *among* teams, to communicate key elements of the overall product. XP can be used *within* teams, which will quickly shake out to form more than a dozen groups, in the case of the largest team, or around four groups, in the case of 30 developers.

One of the chief criticisms of XP is that it does not scale up. For a group of 300, or even 30, XP does not seem to be the choice. However, a quick poll of practitioners reveals that they interact daily with an average of eight teammates. A typical XP team is 6 to 10. Therefore, size hardly matters in choosing among the Spiral Model, UP, and XP.

We demonstrate here the decisions required to choose among the three development methods for an actual project, albeit one that started before any of the methods existed.

At one time, there was a big aerospace contractor that specialized in fixed-wing aircraft. It obtained a contract for a helicopter and put some of its best engineers on the project. Everything seemed the same as previous business, except now the "wings" would be spinning. There were some strict requirements, some of which were immutable, like gross weight, 7,500 pounds, and flyaway cost, $7,500,000. Seemingly, to maintain this theme of sevens, seven processors were required, none being so close together that a single anti-aircraft round could disable the aircraft. Six were used for various parts of the system; the seventh was a "hot spare," meaning that it was powered up, but empty except for the operating system. The remainder of the code was on several nonvolatile (core!) memories hidden in the bowels of the ship.

This system was dedicated to software for navigation and weapons delivery. Everyone on the project thought that this would be relatively simple, since they had done such a system before. How different could navigation be between a fixed and rotary-winged aircraft? How different could weapons delivery be? It turned out to be quite different. None of the weapons were the same, and the helicopter flew lower and slower than any military fixed-wing aircraft. In addition, congressional representatives with oversight of this helicopter kept changing from two pilots to one and back again, while insisting on the same weight and cost—which meant constant redesigning.

Once it was clear that the engineers knew only a little about this type of aircraft, a Spiral Model approach was thought to be the best, so that engineers could start with some key but largely unknown components and then calibrate budget and estimation. Even with the Spiral Model, deliveries were too far between to address congressional changes. UP was chosen, as it consists of iterations within phases.

Concurrent with this could be a way of learning something new about the system with a small team using XP, chosen largely for its speed. The project team initially had to build software for a standard government processor that had words 16 bits in length. Due to the long development time, nobody expected this processor to remain the same. Even government foot-dragging could not cause any change to Moore's Law (processor power doubles every 18 months). Therefore, a small team worked on porting the software to a 32-bit machine, and doing occasional prototypes. XP is well suited to what that group would be doing.

To recap, the team was relatively ignorant of helicopters, so the Spiral Model approach would help build a kernel of the software and calibrate previously successful estimation methods. Each spiral would consist of a UP lifecycle. UP was chosen since it represents an entire software production cycle each spiral, and is iterative itself. The parallel development, consisting of products that would never be delivered, was an XP project. This way, the team could react quickly to requirements changes, like one to two pilots.

One thing on this project was known: it would last a long time—years, maybe decades. This almost certainly eliminates XP from consideration for the core software. The core software development would need heavy documentation to be passed to different developers.

TASK

Which life cycle method would you use for a nuclear reactor controller? Why?

TO REQUIRE OR NOT TO REQUIRE . . . METHODS

Methods have a, well, method of taking over your life. Soon you are concentrating on artifacts since they are required in some way, instead of concentrating on implementing the product. This is why programmers like XP: less weight is given to prose products and more to code.

One very good part of the Team Software Process (TSP) is the Launch. In addition, another two of the pros for requiring it are the weekly status meeting and a tracking tool. These all help start the project and keep it going. In fact, one of the most difficult aspects of a software engineering project is getting started. Some managers believe that getting started is everything, so they impose a method from above. In fact, TSP is so good at encouraging the production of artifacts (that are not code) that it is often imposed. It is easy to confuse the production of artifacts with the production of the product.

There are several reactions possible to the imposition of a certain method. The most serious of these is when engineers are intelligent and reject the method as a result of ego (see Chapter 3). They may perceive any method as too confining and refuse to produce deliverables, preferring to keep procedures in their head. Tomayko noticed this when he worked a lot with configuration management [Tomayko 02]. The theory was that no programmer would make any changes to anything until after the Configuration Control Board (CCB) authorized it. At first, the bright programmers anticipated what would be approved, so they could secretly begin work on the changes. Eventually, they did not wait for the CCB to meet; just kept making changes to defects as they came up. Changes to documents were ignored.

In fact, Watts Humphrey of the SEI once said, "Your real process is the one you follow once you are behind schedule." This means that no matter what method is being used, it goes out the window once programmers perceive the schedule is getting away from them. An effective way of saving time on documentation artifacts is not to start them in the first place. This effect is primarily the reason why some code is costly to repair or otherwise change later: it lacks up-to-date documentation and must be reverse engineered.

This is why we believe in lighter-weight methods with fewer artifacts. The required items are so light that there is time to actually accomplish them. They can also be done at the appropriate time. The design, say, is described near the end of the project, when further changes to it are unlikely. The major difference between established methods and lightweight methods is that in heavyweight methods, artifacts are due at times and in a form that requires them to be changed. Lightweight methods require fewer artifacts, and requires them when what they are documenting is complete, so fewer or no changes are required. This makes programmers quite happy. They find themselves wasting little or no effort. Actually, in a world of "better, faster, cheaper," managers like this format, too. Therefore, if any method is being required, an unobtrusive method is best. Engineers like those best because they seem to require the least work.

TASK

List deliverables from a heavyweight method. Mark the ones needed for maintenance and specify at what life-cycle phase they can be written.

SUMMARY QUESTIONS

1. Compare the three methods discussed in this chapter: What is common among the three methods? What distinguishes them from each other?
2. There are several inherent problems in software development. If you are not familiar with them, just search the Web with a phrase like "problems with software development." Select the five problems that in your opinion are the most critical and explain how each method discussed in this chapter helps solve them.
3. In his paper "The Inevitable Pain of Software Development: Why There Is No Silver Bullet," Daniel Berry [Berry02] argues that although each software development method provides the programmer a way to manage complexity, each method has a catch, a fatal flaw, a task that programmers put off in their haste to get the software done or to do other things. In your

opinion, how does this statement relate to each of the three software development methods presented in this chapter?
4. Programmers can work in different physical environments. The agile methods specifically address the physical aspect of development environments.
   a. Explore different development environments and analyze their fitness for different software development activities.
   b. Analyze your own development environment. What type of work environment are you in? What would you change about it? About how much would the change cost?

FOR FURTHER REVIEW

What would be the problems, if any, in adopting XP to your current workplace. What use would you have to make of the four XP values to help your coworkers make the change?

REFERENCES

[Beck00] Beck, Kent, *Extreme Programming Explained*, Addison-Wesley, 2000.

[Berry02] Berry, Daniel M., "The Inevitable Pain of Software Development: Why There Is No Silver Bullet," *International Workshop on Time-Constrained Requirements Engineering*, 2002.

[Boehm88] Boehm, Barry, "A Spiral Model of Software Development and Enhancement," *IEEE Computer* (May 1988): pp. 61–72.

[Brooks95] Brooks, Frederick P., *Mythical Man-Month*, Second Edition, Addison-Wesley, 1995.

[Brooks87] Brooks, Frederick P., "No Silver Bullet," *Computer* 20(4), (1987): pp. 10–19.

[Jacobson99] Jacobson, Ivar, et al., *The Unified Development Process*, Addison-Wesley, Boston, 1999.

[Mindell02] Mindell, David A., *Between Human and Machine*, Baltimore, Johns Hopkins University Press, 2002.

[Williams and Kessler02] Williams, Laurie, and Kessler, Robert, *Pair Programming Illuminated*, Addison-Wesley, 2002.

[Tomayko02] Tomayko, James, "A Comparison of Pair Programming to Inspections for Software Defect Reduction," *Computer Science Education* (September 2002): pp. 213–222.

ENDNOTES

[1] "Velocity" is like "ideal engineering time." It is the total time you can spend each day on the project, minus breaks, pep talks, and other irrelevancies. For example, if we normally work on a project six hours a day, and two hours disappear somewhere, the velocity is .75.

3 Working in Teams

In This Chapter

- Introduction
- Objectives
- Study Questions
- Relevance for Software Engineering
- Types and Structures of Software Development Teams
- Forming and Rewarding Student Teams
- A Game Theory Perspective of Teamwork
- Outsourcing
- The Bozo Effect
- Mind the Gap
- Summary Questions

INTRODUCTION

This chapter deals with teamwork—one of the main characteristics of software development. Teamwork is essential for the development of any sizable software. Specifically, this chapter addresses types of team structures and dilemmas (whose source is in teamwork) that software developers may face during the process of software development.

The first part of the chapter addresses three types of software teams: democratic, hierarchical, and virtual. Each type is discussed from the perspectives of what kinds of team interaction it inspires, what problems it may raise, and in which situations it fits. The second part of this chapter is dedicated to one issue—rewards—a topic that may raise dilemmas between one's personal interests and benefits one may gain from one's contribution to the teamwork. Like other chapters of this book, this chapter delivers the message that as soon as we increase our awareness with respect to various aspects of software development and various approaches to deal with dilemmas that our profession raises, the better we may cope.

To widen readers' perspective on the topic of teamwork from a theoretical point of view, the chapter ends with a brief explanation of how teamwork in software engineering can be analyzed from a game theory point of view. In addition, issues related to outsourcing in the software industry are discussed.

OBJECTIVES

- Readers will be able to differentiate between democratic and hierarchically based teams with respect to structure, benefits, faults, kinds of projects each structure fits, and roles of team members.
- Readers will become familiar with the roles of typical eXtreme Programming (XP) and Team Software Process (TSP) team members.
- Readers will analyze benefits and pitfalls of different kinds of teams such as "jelled" teams and virtual teams.
- Readers will be able to organize an appropriate team.
- Readers will understand how incentives work and influence software teams.
- Readers will understand what outsourcing means in the software industry.

STUDY QUESTIONS

1. What software development methods use democratic teams?
2. What software development methods use hierarchical teams?
3. What type of team makes you feel the most comfortable, and why?
4. What considerations should one address when constructing a software team?
5. As a software project leader, what roles would you assign to team members?
6. Where is delegation evident in software development?
7. In your opinion, would software developers prefer equal allocation of bonuses among team members? Explain your opinion.
8. How does outsourcing work in the software industry?

RELEVANCE FOR SOFTWARE ENGINEERING

This chapter is about software development teams: how they are structured, how they encourage communication, their roles, and some of their problems. As mentioned previously, software engineers spend most of their working life as members of a team, so it is important that they know why they are organized in a certain way, what the nature of their role is, and how to fit in well with other team members. Software engineers are likely to have others from different nationalities and cultures on their team, and they should know how to act toward them as well.

As it turns out, cooperation is an essential attribute of the process that guides the development of software. At the same time, it may raise some conflicts between one's wish to excel and one's need to contribute to the teamwork. Thus, it is important that software engineers be familiar with dilemmas that this need for cooperation may raise and with different approaches to cope with such dilemmas.

TYPES AND STRUCTURES OF SOFTWARE DEVELOPMENT TEAMS

The three types of teams that we discuss in this chapter are democratic, hierarchical, and virtual. It is true that virtual teams can be organized either democratically or hierarchically, but this type is of such growing prevalence that we treat it separately. To highlight the idea that the creation of a software team should not be random, but rather based on some kind of analysis, several factors (for example, the development method used and the teammates' level of experience) on which team structure is dependent are emphasized.

Democratic Teams

Democratic teams are characterized by everyone being a peer, or equal in another way. Student teams are often democratic, since peerism is more obvious. The organizational structure of a democratic team is a regular polygon (see Figure 3.1). Each programmer is an equal member of this circle; the decisions a programmer makes or the opinions a programmer holds are considered of weight equal to the others.

Obviously, beginners like democratic teams, because they are often treated like more experienced team members. This is patently impractical, so a "virtual" hierarchy is set up, where the beginner often does not go against senior engineers. Another way to keep the advantages of a democratic team but not lose the effects of seniority is to put the beginner in a pair with a more experienced engineer (see Chapter 2, "Software Engineering Methods"). This way, the pair can establish a hierarchy together, yet appear and act democratically in a group.

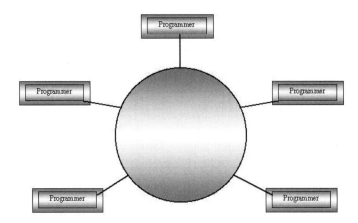

FIGURE 3.1 A democratic organization.

Engineers will form democratic teams out of nearly any method. In some ways, the Team Software Process (TSP) is the most democratic [Humphrey00]. Each role in the process has appended "Manager," as in "Development Manager." Humphrey hardly assumes each member has equal experience (example tasking is Development Manager, Planning Manager, Quality Manager, Process Manager, and Project Manager). A Development Manager decides on the number of iterations a team does in building software and develops a strategy to produce the product. The Planning Manager takes charge when iterations are planned, to assist the team in estimating and load balancing. The Quality Manager produces things like the configuration management plan and the acceptance tests. A Process Manager is like a Coach in an agile process, having the answer to most questions about TSP. The Project Manager has fewer responsibilities than on a non-TSP project, as the other managers do some tasks that lie in traditional methods. However, finding resources, running the meetings, and risk management are tasks still under the Project Manager's purview.

If a team has five or fewer members, then this TSP approach lends itself to a democratic organization. Otherwise, it lends itself to a hierarchical approach, as we discuss in the next section.

TASK

How is the code development allocated to TSP team members?

One time when everyone has roughly equal experience is in school. No matter which method the instructors or the team uses, it quickly takes on the aspects of a democratic method when used in an academic environment. This tendency to seek out democratic organizations is counterintuitive for some methods, such as those used for large-scale projects. However, the fact that they are sought out demonstrates student desires. Reward techniques may also affect team organization (see *Forming and Rewarding Student Teams* in this chapter).

The biggest problem with democratic teams is that no one is in charge. Actually, we should say everyone is in charge, and no one is responsible. Anyone wanting a decision from a democratic team must wait for it to achieve consensus. This may take some time.

TASKS

1. Suggest pros and cons of the last described phenomenon (that decision-making processes of democratic teams may be long).
2. How can a decision-making process be enhanced in democratic teams? Suggest at least two mechanisms.

Another advantage to a democratic team is ease of replacement. Any engineer can be replaced by nearly anyone. The democratic team has a competence measured by the sum of its parts. Therefore, the loss of a member does not change the overall competence much. A further positive to a democratic team is the rapidity of "jelling." A jelled team is one that finds working with others a more positive experience than working alone. It is when the team becomes greater than the sum of its parts. Many software managers emphasize jelling. A team that works closely together is easier to deploy.

TASK

What problems can a democratic team structure raise?

A team using agile processes probably functions better as a democratic team. There are no layers between the manager and the engineers, and communication, valued by many agile processes, is enhanced. Looking at agility from the perspective of team structure, a democratic team fits agile processes better. "Collective ownership" is an XP practice that ensures the democratic approach. As a practice, collective ownership means that anyone on the team can change the code. Obviously, this is difficult to set up in a hierarchy. A large amount of trust and a small amount of ego are necessary for this to work.

Actually, we think a small ego is necessary for any democratic team. Otherwise, consensus takes forever to achieve, and frustrated team members act independently. Chaos can result. However, small egos are seemingly difficult to come by in the software business. Many programmers are evaluated by individual achievement. These people often are too selfish to be suited for democratic organizations. This issue is further covered in this chapter under the discussion about rewards in software teams. Still, most XP teams operate democratically. This is due to collective ownership and to the nature of agile processes in general.

TASKS

1. Based on Chapter 2, what additional XP practices may support the creation of gelled teams?
2. What stages in software development are appropriate to be managed using a democratic organization?

Hierarchical Teams

Hierarchical teams are organized like a tree (see Figure 3.2).

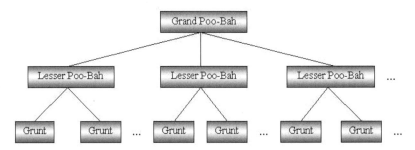

FIGURE 3.2 A hierarchical organization.

Hierarchies are common. Many groups use such an organization outside of software development. Certainly, Mills' Chief Programmer Team [Mills83] and Brooks' Surgical Team [Brooks95] are software development groups using this organization. It is very easy to scale up hierarchical teams—one can just add a layer. Unfortunately, this hampers communication, requiring the overall manager to cross many layers to get to the workers.

Delegation is part of a hierarchical management. It can be accomplished in a positive way by preparing some engineers to move to management. It can also be done in a negative way by passing unwanted tasks down the hierarchy.

Self-esteem is often more difficult to obtain in hierarchies. Depending on the number of layers, the Grunts (low-level programmers) may feel powerless. Moreover, layers create distance, which makes the Grunts feel they are not really contributing to the company. This is especially true if two or more layers are in a different geographical location. At the lowest level, there is more loyalty to the team than to the company. The home company is often an abstraction of the same sort as the competition. Besides, Grunts in such team organizations may miss the big picture of what goes on and may have only a narrow understanding of the development environment in general and the developed software in particular. This fact may influence decisions they make during the development process.

TASK

How can a company increase its employees' loyalty when two or more layers are in different geographic locations?

Additionally, in hierarchical teams, the manager has to be respected, not necessarily liked. Otherwise, few workers would make the sacrifices necessary to follow in modern workplaces. There are five sources of power, in order of increasing desirability: punishment, reward, legitimate, identification, and expert.

The least effective of these, punishment and reward, are the way humans train dogs. Punishment, at the lowest level of power, can be either physical or psychic. If physical, like hitting a dog with a paper (good luck later training the dog to get the paper!), pain is caused by the manager whenever the employee does something judged "wrong." Unless employees are somehow constrained, they will not stay long in this type of atmosphere. Punishment is like speaking sharply to a dog. Since the dog is genetically primed to please humans, a rebuke is like striking it. Oddly enough, some people, especially those whose self-esteem is wrapped up in pleasing elders and bosses, put up with this kind of treatment for years. Nevertheless, it is the weakest source of power.

Next highest is reward. If either a dog or a human gets some positive reinforcement for doing something "right," we are using reward. Both dogs and humans are affected by rewards, as simple as a treat for a dog to monetary rewards for a human. Pavlov's dog and Operant Conditioning prove these points. This is where many companies stop. They might have a variety of rewards, and that is it. Dogs cannot go any further along this scale, due to a lack of intelligence. In the continuation of this chapter, we discuss connections between rewards and student teams in the university. However, in that context, rewards are used to raise dilemmas in software development and not as a means for gaining power or respect.

The next most effective source of power is "legitimate power." This is like a replacement in a military unit. If the replacement is a common soldier, then the replacement fits into the unit without much fanfare. If it is a new platoon leader, then the replacement is obeyed like the former platoon leader. This works for a while, due to legitimate power, since officers are in charge of troops (note that this is easier in a hierarchical team). However, if an officer regresses to the level of punishment or reward, the officer quickly loses the advantages of legitimacy. In a software, or other, team, a "manager" is considered "in charge" when appointed. Thus, we have the "honeymoon period," while the new leader's personality comes out.

The next highest source of power is "identification." This is the source used by the most liked managers. Workers follow these managers because they can identify with their style and personality. They want to be like the manager some day. These types of managers seems relaxed with their power. They are easier to get along with, but there is no doubt who is the boss.

TASK

Identify five characteristics of a team leader with which software developers may identify. Why are these characteristics important in the case of software development?

The final source of power is "expert." This is not necessarily a more effective source than identification, but it is powerful. Experts are very smart and may have 1,000 ideas a day. They are also very quick to think, jumping from topic to topic, and are very unpredictable. Of their 1,000 ideas, 999 are garbage, but one is worth seven figures. The problem is that someone, usually a slower, less intelligent underling, has to go through the entire list trying to find the valuable one. The next day there are another 1,000 ideas. The only reason people put up with this is the expert source of power. These managers are so obviously brilliant and a cut above even the intelligent people around them, that people follow the managers like puppies. When they do not want to follow, it is easier to work around or anticipate the moves of the manager rather than depose this manager from the current position.

TASK

Identify five problems with a software team leader who is an "expert."

A larger team is able to use TSP to overcome the vagaries of single leadership. It may combine the effort distribution of TSP and democratic teams with some advantages of hierarchical teams. The Project Manager has some new responsibilities. Among other activities, the Project Manager must now communicate for the team, insulate the members, and generally represent the team. The other members can delegate parts of their tasks, although everyone still estimates.

We said that bigger TSP teams could lend themselves to a hierarchical approach. This means that under every "manager" are one to many Grunts. A Project Manager can have the other managers as underlings in a purely hierarchical team, thus fulfilling the three layers depicted previously. Otherwise, a hybrid team can be formed: democratic at the named manager level, hierarchical below. Delegation techniques are necessary for this type of team. The objective of delegation ought to be training of the Grunts to be Lesser Pooh-Bahs and so on up the scale. There is, of course, the motive of getting things done in parallel. For example, the Grunts assigned to the Process Manager can each be assigned one or more subteams to observe their process, and flag any deficiencies. The Process Manager can observe the other Managers for the same reason.

Often, there is a reluctance to delegate. This reluctance can have several origins. One is that people are afraid to appear replaceable. This is a big problem, because some younger members worry about being cast aside. Another disincentive is the belief that "it is only done well if you do it yourself." This is generally evidenced by the delegator's spending time cajoling the delegate. The reason is subtly different from appearing incompetent as a person; often the team appears incompetent as a group.

Ego size is a factor in the success of hierarchical teams. If Grunts are "seen and not heard," valuable feedback is lessened. Grunts or any other underlings need a significant ego to speak up in a crowd. However, even though a somewhat larger ego is needed for communication, if the ego is too big, following directives from above is problematic. If someone with a large ego is saddled with a superior using "reward" measures to lead, the egotist will see right through his superior and ignore his authority. This causes -low-level cooperation to fail, and makes a subteam essentially leaderless.

Mills and Brooks have an answer to this. Mills' "Chief Programmer" team organization [Mills83] seems typically hierarchical. The team is small, consisting of only three members: the Chief Programmer, the Backup Programmer, and the Librarian. There is a very strict pecking order: Librarian, Backup, and Chief, from bottom to top. The Librarians used to punch cards, keep them in order, maintain version control, and pick up and distribute output. They still maintain version control and are expert in Integrated Development Environments (IDEs, like Code Warrior). The Backup Programmer does everything the Chief Programmer does, but is sort of a Boswell. The Chief Programmer is more experienced and obviously in charge. The Chief Programmer is responsible for the architecture, and most of the nonroutine code. The routine or reused code is done by the Backup. Thus, the Chief can use the Backup to bounce ideas off, but remains responsible for the decisions.

Brooks [Brooks95] expanded the Chief Programmer team to become the "Surgical Team" and encompass 10 persons. These include the familiar Chief Programmer, Backup Programmer, and Librarian. Brooks added an editor, two secretaries, a tool smith, an administrator, a tester, and a language lawyer. The editor is for documentation produced by the team; the secretaries are divided between editor and administrator for routine tasks such as setting up travel. The administrator is to keep ordering items and other tedious nonproduct-related tasks from the Chief, enabling that person to concentrate on what he or she does best. The toolsmith knows more about the operating system and tools than anyone else does. The tester does functional testing. The language lawyer seemingly knows more about the language chosen than anyone on the planet[1]. Obviously, these seven additions are underlings. They may even be only part-time group members.

TASK

Identify benefits and disadvantages of a Surgical Team.

How is a hierarchical team scaled up? Add several Surgical Teams together to produce the code for each component. This way, we keep subteams relatively small

(10), with the entire team being of indefinite size. People report that they can handle interactions with about eight others. If you eliminate the secretaries in a Surgical Team, you are left with eight people.

TASKS

1. Why does the number of team members matter? What are the main problems with too-big teams? With too-small teams?
2. Identify and discuss instances of hierarchies we encounter in everyday life. What can we learn from these hierarchies about the working and management of hierarchical software teams?
3. What is a downside to delegation? Discuss.

Discussion

1. Why are many people comfortable with democratic teams?
2. For what types of software are democratic and hierarchical teams useful? For what types of software are democratic and hierarchical teams not useful?

Virtual Teams

Virtual teams can have either a democratic or a hierarchical organization. In a virtual team many of the team members are not co-located physically. For example, part of the team can be in Bangalore, India, another part in Palo Alto, California, and another in Haifa, Israel. A team like this has roughly an hour of work time overlap between Bangalore and Palo Alto, and later between Palo Alto and Haifa.

The adoption of these types of teams is mostly inspired by cost. Indian programmers are paid only about a quarter what Americans are paid, but even their managers worry about outsourcing to Thailand or China, where programmers are paid even less! Typical gross savings rates are between 50 percent and 80 percent. Due to the more elaborate overhead of many virtual teams, U.S. companies realize a net savings of only 15 percent to 25 percent. Nevertheless, 80 percent of IT shops are discussing outsourcing in 2004 [King03].

Discussion

What ethical and cultural problems may members of a virtual team face?

Running a distant second as a reason for adopting virtual teams nowadays is their more efficient use of equipment. When computer centers were anchored by mainframes, it was more sensible to pass their use to others in a different time zone once most of the workers in one time zone went home. Now, most computing is done by desktops, which are about the same price wherever you get them.

Even though many virtual teams are now formed to save money, a person who is a member or leader of a virtual team has many possible problems. These include distance, which means less context, and cultural inequalities.

Cultural difficulties can include such problems as members being unable to operate in a democratic team, making consensus difficult. Many are trained to be solicitous of Pooh-Bahs, and take every wish as a command. They are not used to giving, or getting, negative feedback.

Virtual teams also can use some relatively ponderous methods. For example, many virtual subteams work from a "finished" statement of work, like a waterfall software life cycle. They negotiate based on requirements, so any changes are far-reaching and a problem. Members of the agile software development community claim that they code faster (thus, less expensively) and handle change, claiming that they obviate the need for outsourcing offshore.

Moreover, it is difficult to measure outsourced progress. Virtual teams can hide behind documentation. They can produce design documents or architectures that are massive in scope and give the appearance of progress. These documents have to be useful, no matter how far apart the subteams are, which means the documents must be short and to the point. No one will read a 330-page document.

Remote outsourcers usually begin work once the requirements are settled. They can stall code production by hiding behind documentation, as described previously. Frequently, an absent client will see releases so rarely they cannot tell whether to cancel a late one. Basically, then, agile processes are difficult, if not impossible to install in a virtual team, due to the need for co-location.

Despite their drawbacks, virtual teams appear to be the wave of the future, mostly for cost reasons. Still, some investment is necessary. It is often better to have one or more team members cross oceans to be present at the first meeting so that during later meetings, held using shared desktops and Voice over IP (VoIP), the widely separated members can visualize each other. The strength of this type of live contact cannot be stressed enough for building a gelled team at a distance. It turns out that outsourcing to Ireland or Northern Ireland results in a travel expense of less than $1 per hour per team member, with the savings from going off-shore built in. With travel costs as low as these, jelling is facilitated.

In Chapter 6, "International Perspective on Software Engineering," the discussion of software engineering from an international perspective is widened.

FORMING AND REWARDING STUDENT TEAMS

The question of how to allocate incentives, rewards, and bonuses among team members is relevant with respect to many professionals and kinds of institutions.

Reward allocation with respect to software engineering is important for at least three reasons:

1. Teamwork is essential for software development. As a result, conflicts between the contribution to the teamwork and the way in which rewards are shared may intensify.
2. Management of software development teams and communication among team members are complex issues.
3. Software developers are usually highly motivated. Their motivation can cause conflicts between personal targets and team goals.

Many people using this book may be in an academic department where team composition and grading are problematic. Thus, the discussion in the continuation of this chapter is about software projects that are developed in teams in the framework of university undergraduate courses, although some results are applicable to practitioners. Specifically, this section focuses on the conceptions of software engineering students of the relationships between reward and cooperation in the context of software development, and it is based on students' ways of coping with a conflict between their urge to express personal skills and the unavoidable need to cooperate with their teammates. In order to disassociate the grade issue (and make this more relevant to practitioners), the discussion is conducted in the context of financial reward allocation offered when working on a software project.

The discussion is based on the following activity, conducted in three stages. (For the full report, see [Hazzan03]).

Activity (Explained after Step 3)

Step 1: Individual Work

1. Assume that you are a member of a software development team. Your team is told that if the project it is working on is successfully completed on time, the team will receive a bonus. Five options for bonus allocation are outlined in Table 3.1. Explain how each option might influence team cooperation and select the option you prefer.

Step 2: Team Work

1. Each team decides on one option that it prefers.

Step 3: Individual Work—Reaction to Two Situations

1. Your team leader tells each team member, separately, that personal performance and achievements are major factors contributing toward promo-

TABLE 3.1 Bonus Allocation Worksheet

	Personal Bonus (% of the total bonus)	Team Bonus (% of the total bonus)	How this option may influence teammates' cooperation
a	100	0	
b	80	20	
c	50	50	
d	20	80	
e	0	100	

tion. The team members do not know that others have been told the same thing.
   a) How will this affect team collaboration?
   b) As a team member, how would you suggest sharing the bonus now?
   c) How would you behave in such a situation?
3. Now, your team leader tells each team member, separately, that contribution to teamwork is a major factor contributing toward promotion. The team members do not know the others have been told the same.
   a) How will this affect team collaboration?
   b) As a team member, how would you suggest sharing the bonus now?
   c) How would you behave in such a situation?

Explanation of the Task

Step 1 focuses on students' preferences when a neutral situation is described. Step 2 examines how students face possible conflicts between their own preferences and the preferences of their team members. Before proceeding with Step 2, students' written responses to Step 1 are collected to ensure that the answers to Step 1 are not changed later.

Step 3 presents the students with two cases. The first addresses a situation in which students have a personal incentive. The second describes a situation in which there is an incentive to contribute to the teamwork. In both cases, personal promotion is conditional. In the first situation, if a student contributes to the teamwork, the student's personal contribution might not be outstanding and thus the student might not be promoted. However, if the student does not contribute to the teamwork, the project may fail and his or her personal contribution might not be relevant. Thus, students are faced with a dilemma whether to contribute to the teamwork or not. In addition, they should decide how they prefer the bonus to be shared in this case.

In the second situation, there is a high incentive to contribute to the teamwork. In this case, it would be logical to contribute to the teamwork. As a result, everyone involved will benefit: the project will be completed on time, and all the team members will get their bonus. The dilemma in this case is faced by team members who in Step 1 preferred that a higher ratio of the total bonus be divided on an individual basis. Such teammates might face the following conflict: on the one hand, they prefer that a higher portion of the total bonus be divided based on personal contribution and, as a result, their contribution to the teamwork might tend to be dominated by their personal-level contributions. On the other hand, this situation (the second of Step 3) encourages teamwork. In other words, teammates might face a dilemma between being dominant (and receiving a higher portion of the total bonus) and contributing to the teamwork (in order to be promoted, as promised by the team leader in this situation).

Discussion

The different stages of the preceding activity raise many questions, emotions, and dilemmas with respect to software teams. Discuss these issues of software teamwork.

According to the results reported in [Hazzan03], the majority of the students prefer that a small portion of the bonus be allocated based on individual contributions, while the majority of the reward is divided equally among the team members. Specifically, most of the students prefer the option according to which 80 percent of the bonus is distributed equally among team members, and 20 percent is divided according to the personal contributions of each team member.

In what follows, we examine student preferences and tendencies to change their preferences when conditions change (in Step 3). As a basis for this analysis, we focus on data obtained from two classes.

Class A: In the neutral situation (Step 1) and in the individual-based incentive case (Step 3, Question 1), students prefer 80 percent of the bonus to be divided equally among the team members and 20 percent of it to be allocated on a personal basis. The majority of the students preferred the opposite option of bonus allocation when a team-based incentive is involved (Step 3, Question 2).

It was clearly observed that all Class A members either increased or did not change the team bonus when an individual-based incentive was involved, and that all of them increased or did not change the personal bonus when a team-based incentive was introduced.

In the class discussion that took place following the activity, the students in our sample explained that they increased the *team* bonus when there is an incentive to increase the *personal* bonus (Step 3, Question 1) because the promotion is conceived as compensation. In other words, they view the promotion as an alternative reward. Thus, knowing that they will be promoted, they tend to be generous with their teammates and increase the team bonus. The fact that they increased the *personal* bonus when there is an incentive to increase the *team* bonus (Step 3, Question 2) is explained by the students as follows: on the one hand, the promised promotion encourages them to contribute to the teamwork. Furthermore, students were fully aware of the fact that software development is based on teamwork and that cooperation is essential to this process. On the other hand, the increased individual-based bonus that they suggested in this case will encourage them to excel on the individual level.

Class B: Responses of students from Class B reflect a different picture. No significant preference is predominant in any of the steps. In general, students do not change their preferences when the conditions change. Second, when they do change their preferences, they tend to increase the personal bonus when an individual-based incentive exists. When there is a team-based incentive, Class B students tend to increase the team bonus.

TASK

Suggest additional attitudes to the various situations described in the activity. How can you explain them?

This variety of patterns suggests that perhaps different kinds of teams should be evaluated in different ways. There are teams for which the individual-based incentive is better suited, and there are teams for which the team-based incentive is more appropriate. For example, Class A, whose members tend to *decrease* the personal bonus when there is a personal-based incentive (because the promotion is conceived as compensation), may be characterized as team oriented. However, Class B, whose members tend to *increase* the personal bonus when an individual-based incentive is present, may be characterized as individual oriented.

Furthermore, some students declare that their preference of bonus allocation is determined by their assessment of their personal skills relative to those of the other team members. Thus, if they know that the other team members have more or less the same personal skills, they would tend to reduce the percentage of the personal portion of the reward. However, if they assess their personal skills as being higher than those of the other team members, they will increase the personal portion of the reward.

We end this section by briefly discussing the issue of forming teams and evaluating student projects.

Forming Student Teams

Should we form teams of students with similar preferences, with different preferences, or according to a specific combination of preferences? Since it was found in Step 1 that a majority of the students in four of the five classes (that participated in the research) chose to allocate at least 80 percent of the reward on a team basis, it might make sense to form a team composed of a majority of such students, balanced by some students who prefer to allocate less than 80 percent on a team basis. The idea behind this suggestion is to create a team in which the majority of its members have similar beliefs regarding teamwork, and to add to it students who are more extreme in their preferences. Thus, extreme individual tendencies are eliminated, and the individuals who have slightly different perspectives will have to adjust, at least partially, their personal preferences with respect to bonus allocation.

Other opinions, such as presented in [Redmond01], suggest that students should form groups, instead of being assigned to a team.

TASKS

1. Is there a difference between forming student teams and forming software teams in the industry? Should other criteria be considered?
2. Based on the presented data, suggest other approaches to be implemented in the process of forming student teams and in forming software teams in the software industry.

Evaluation of Student Projects

The previously presented observations can also be used to guide the decision on the evaluation policy to be used. Accordingly, teams that tend to cooperate will be rewarded in a more equal manner, and vice versa. The implications of such a method might be that students will be more satisfied with the grading policy (since it is compatible with their belief system) and will tend to invest more efforts in their projects.

The preceding conclusion stands in line with some of the lessons suggested by [Brown95], two of which are relevant to the discussion here:

- Recognize that different types of teams and teamwork require different types of reward schemes.
- Realize that individual and team incentives are often complementary, not conflicting, components of your reward strategy.

Section Summary

This section examines the theme of reward allocation. It is based on the reactions of students, whose preferences can be used as a tool in the determination of the grading policy of students' software projects. This experiment lends itself to expansion to real-life situations, with real companies offering real financial rewards. Within this broader setting, such a study can be used to determine how and under what conditions a specific method of reward allocation leads to greater team cooperation. Another interesting question to explore is software developers' preferences in different countries and cultures.

A GAME THEORY PERSPECTIVE OF TEAMWORK

We end this chapter by highlighting the topic of software teamwork from a game theory perspective. Generally speaking, game theory analyzes human behavior and uses for this purpose different theoretical fields such as mathematics, economics, and other social and behavioral sciences. Game theory is concerned with how individuals make decisions when they are mutually interdependent and attempts to describe how people interact in their attempt to win in such cases. In game theory situations, the participants want to maximize their profit by choosing particular courses of action, and that each player's final profit depends on the courses of action chosen by the other players.

TASK

Review *www.gametheory.net* or another comprehensive game theory Web site for further information about game theory.

- Summarize five lessons you learned from this review.
- Suggest connections or implications of game theory to software engineering in general and to software teams in particular.

Within game theory, we focus in this chapter on one framework: the Prisoner's Dilemma. This framework illustrates why people tend to compete even in situations in which they might gain more from cooperation. This behavior is explained by lack of trust. In its simplest form, in each turn of the game each of two players has a choice between cooperation and competition. The working assumption is that each player does not know how the other player behaves and the two players are unable to communicate. Depending on the two players' decisions, each player gains

TABLE 3.2 The Prisoner's Dilemma

Action of A \ Action of B	Cooperate	Compete
Cooperate	+5	−10
Compete	+10	−5

points according to the payoff matrix presented in Table 3.2, which describes the dilemma from A's perspective.

The values in Table 3.2 are illustrative. However, they do indicate the relative benefits one gains from each selection. Accordingly, the table indicates that when a player does not know how the other player behaves, it is worth choosing competition. In other words, if A does not know how B behaves, in both cases (whether B competes or cooperates) A will get more if it chooses competition. However, as can be seen, if both choose the same behavior, they will gain more if they cooperate rather than if they compete.

Game theory is named for police scenarios in which two partners in crime are placed in separate rooms at the police station. Each is offered a similar deal; if the first implicates the other, the first may go free while the second receives life in prison. If neither implicates the other, both are given moderate punishment, and if both implicate the other, the punishment for both is severe. They have to make a decision without knowing their partner's choice.

As it turns out, such cases happen in many real-life situations in which people tend to compete instead of cooperate, when they might gain more from cooperation. The fact that people tend not to cooperate is explained by people's worries that their cooperation will not be reciprocated and they will lose more (than if they compete and their partners cooperate). As has been indicated, this happens (in most cases) because of a lack of trust between partners.

TASK

Visit the *http://www.gametheory.net/* Web site. Select one of the Java™ applets that illustrate the Prisoner's Dilemma and play with it. What was your strategy? Did you win? What did you learn from this game?

In what follows we use the Prisoner's Dilemma to illustrate that the source of some of the problems in software development is competition among team members. For this aim, let us first observe that cooperation (and competition) can be expressed in software development environments in different ways, such as information sharing (versus information hiding), using coding standards (versus not following these standards), writing clear and simple code (versus developing a complicated code), and so forth. It is reasonable to assume that such expressions of cooperation lead in most cases to a project's success, while such expressions of competition may lead to a project's failure.

Because cooperation is so vital in software engineering processes, it seems that the Prisoner's Dilemma is strengthened in software development environments. To illustrate this fact, let us revisit the reward issue and examine a story in which a software team is promised that if it completes a project on time, it will get a bonus allocated according to the individual contribution of each team member. To simplify the story, let us assume that the team comprises two developers, and let us construct the payoff table (see Table 3.3).

The difference between Table 3.3 and the original Prisoner's Dilemma table (Table 3.2) lies in the cell in which the two team members compete. In the original table this cell reflects a better outcome for each than the cell in which they cooperate and their partner competes, but in this case, which refers to software development situations, it is the worst for both of them. This fact is explained by the vital need for cooperation in software development processes. As can be observed, in

TABLE 3.3 The Prisoner's Dilemma in Software Development Environments

	B Cooperates	**B Competes**
A Cooperates	The project is completed on time, and A and B get the bonus. Their personal contribution is evaluated as similar and they share the bonus equally: 50% to A; 50% to B	A's cooperation leads to the project's completion on time and the team gets the bonus. However, as A had to dedicate part of the development time to understanding the complicated code B wrote, A's personal contribution to the project is evaluated as smaller than B's. As a result, B gets 80% of the bonus and A gets 20% of the bonus
A Competes	A similar analysis as when A cooperates, B competes, but the allocation in this case is reversed: 20% to B; 80% to A	Because both A and B compete, they do not complete the project on time, the project fails, and no one gets a bonus.

software development environments there are cases in which partial cooperation is preferred over no cooperation at all.

As it turns out, in many software development environments, team members are asked to cooperate even when they cannot ensure that their cooperation is returned. In such cases, as the Prisoner's Dilemma table indicates, each team member would prefer to compete. Because the reciprocation of cooperation is unknown, each team member cannot trust the other teammates, and all compete. As Table 3.3 indicates, in software development environments, such behavior gets the worst result for all team members.

Discussion

Discuss possible ways to increase trust among software team members, and how these ways might enhance cooperation among software team members.

TASK

Review the eXtreme Programming practices (see Chapter 2). In your opinion, which of the XP practices may increase trust among team members? Using the Prisoner's Dilemma framework, explain in what ways this trust might lead to cooperation and how the team members will gain more from such cooperation.

OUTSOURCING

With respect to outsourcing, we discuss two related topics: virtual teams and the Capability Maturity Model.

A Common Use for Virtual Teams: Outsourcing

Managers often use virtual teams as part of their outsourcing (moving jobs offshore) strategy. They see that an Indian developer might be paid $5/hour for a job an American developer might get $15/hour to do [Hayes04]. The choice seems clear: any manager who wants to save funds should send the jobs overseas, where their money will go further.

This concentration on savings hides the reality of the situation: outsourcing is not so profitable. Perhaps only 15 percent to 20 percent is saved by the effective companies. Virtual teams need experienced leadership; they also need collaborative tools. Both are still a cost, even if not a large one. Nevertheless, the idea is spreading. Only about 27,000 jobs were shipped overseas in 2000, but the number is expected to be a half-million by 2015. Many of these jobs will be outsourced to India, which has a large English-speaking population, well-motivated workers, and a good educational system for persons of above-average intelligence. Such jobs are beginning to show up in near shore areas of Mexico and South America.

Many companies rushed into having virtual distributed teams because they thought that they would save money or simply because other companies were doing it. Recently, some companies are bringing back home such things as help centers. The main reason for outsourcing—that of giving customers better service—was not accomplished by putting the end of some telephone lines overseas.

Obviously, sending these jobs overseas coincided with the bursting of the dot-com bubble and the inevitable recovery. Growth in outsourcing matches a renewed need for programmers. This, naturally, caused a big backlash among unemployed programmers. How this will all work out is not yet clear, but it is certainly on every company's radar screen.

One final observation: large companies with big IT groups tend to outsource less. We suggest that this is because they already have staff in place, and do not want to suffer morale problems, misplaced costs, or customer dissatisfaction.

TASKS

1. Describe the organization of a distributed virtual team.
2. Where are the most effective places for outsourcing? Why?

The Capability Maturity Model and Its Use for Outsourcing

One generally accepted way of identifying companies with which to outsource is their Capability Maturity Model (CMM) ranking. The thinking is like this: "If we compare Maturity Level rankings, and find a company of equal or higher rank, we can successfully outsource to them." It is not this simple.

The CMM was developed at the Software Engineering Institute (SEI) late in the 1980s and early in the 1990s. Its intent was to measure the capability of part of an organization to develop software. There are also SEI maturity models for other fields, like systems engineering. All posit five maturity levels. Most companies have been identified as Level 1 or 2, with the first Level 5 organization—the Motorola branch in Bangalore, India—found in software. Since then, many Indian companies have tried for some ranking or another. It is not obvious that they are doing this to increase their chances at getting outsourcing contracts, although that has been among the fallout. The SEI quit supporting the CMM at the end of 2003, although there is some effort on the part of Carnegie-Mellon University, which runs the SEI, to keep the CMM around. This would be a welcome development by those already appraised at some level or those already working toward a ranking.

Why is comparing Maturity Levels not so simple? Because the CMM is a measure of *having* something, not of how well an organization *does* something. In other words, an organization can have the worst requirements management system imaginable, but the fact that they have one allows them to check the item off the Level 2 Key Process Area. We observed that a recently assessed near-shore company had many software development life cycle models in their process handbook, and teams in the organization used every one, but there was no guidance as to which life cycle to use in a specific instance.

Therefore, an organization can have a high level of *capability*, but not a high level of accomplishment. Of course, this can play havoc when working together with a company that tried only to achieve a rating.

TASK

Go to the SEI's Web site at *www.sei.cmu.edu* and find out why the integrated version of the various CMMs, the CMMI, is supplanting the CMM.

THE BOZO EFFECT

One important factor of team work is whether the team has "jelled." This refers to when the members of the team work so well together, they are "more than the sum of their parts." In other words, their output is greater than they could have achieved individually, added up.

Whether a team jells is often a function of individual attitudes; specifically, of egos. It is true that programmers need egos. They probably need a big one to maintain an attitude of perfection while solving difficult problems. However, they must find a way to submerge their egos in forming a team or the team will never jell. We have observed the effects of this in action. We refer to effectiveness of teams relative to egos as the "Bozo Effect."

Bozo was a clown character. Sometimes, people who disrupt teams are referred to as "Bozos." The Bozo Effect is a function of how many Bozos can be part of the team and still have it deliver reasonably working software.

Our observation, not surprisingly, is that smaller teams, up to four people, can be totally disrupted by a single Bozo. This is usually a person who leads with ego, and refuses to adhere to the decisions of the group in case of disagreement. Eventually, the tension is so great that the team will self-destruct.

Conversely, a larger team can handle more Bozos. We have observed that up to three Bozos can be part of a team of eight and it will still deliver. This is probably because a majority of the team is still productive. The Bozos are drowned out, and clients are used to late software, unfortunately.

This legislates against the wrong type of "star" performers. Another of our observations, this time of many hundreds of student programmers, is that less than 1 percent are "intuitive" programmers, and the rest are "imitative" programmers. The difference is that the intuitive programmers can code from the solution that seems to be already in their heads. This solution will normally be the correct one, but when you ask intuitive programmers how they arrived at a solution, they are as mystified as you are. Imitative programmers are "reflection in practice" adherents (see Chapter 10, "Learning Processes in Software Engineering"). When they observe an aspect of a problem, they might think, "a loop worked here that time I had a similar problem." They often do not invent solutions; they copy them from past experience.

We deduce that having larger numbers of imitative programmers is not a bad thing. In fact, it seems simpler for imitative programmers to reduce the effects of their egos. Moreover, intuitive programmers tend to not make sufficient documentation. However, intuitive programmers seem to do better on programming tests, and upper management often searches for them, ignoring their shortcomings, because they can be more productive in a shorter period of time.

TASK

Name the tools of the imitative programmer.

MIND THE GAP

Anyone who has ridden on an English train has seen this inexplicable phrase on a sign near an exit door. It means there is a gap between the train and the station's platform. Do not fall in it. There is also a major gap between the best and worst programmers. This gap, if not minded, may make it difficult or impossible to successfully populate a democratic team.

A gap this large is historically quite possible. For example, the best fighter pilots have great shoot-down rates. Others fly every day and shoot down nothing. One German on the Eastern Front during World War II brought down over 350 Soviet airplanes. How was he so productive? He had a process: he flew closer and closer to the enemy's rear until one burst from his guns would make the other airplane explode. In fact, several times, debris from these close explosions brought down his own airplane. Still, comparing this pilot to the nonproductive ones reveals a prodigious gap.

The gap between the best and worst programmer is smaller. There are habits that can close this gap. A devotion to documentation is one. With adequate explanations, a project can be passed on. Those at the top of the heap as programmers rarely have that habit.

Balance is crucial here. If the project is short lived and a one-time job, then using star programmers is generally all right. However, if the source code will be maintained and used for years, programmers of lesser ability, even though slower as developers, are more likely to ensure that the project can move forward.

SUMMARY QUESTIONS

1. Which type of team would you like to work in and why?
2. Interview five software engineers who work as a group in a software house. Identify the team structure of which they are a part. Identify the reasons that led to the establishment of this team structure. In your opinion, is it the appropriate team structure for the project you were told about?

FOR FURTHER REVIEW

1. You are given a rough vision document relating to a software project. Write a team organization for a democratic, hierarchical, and virtual team.
2. Take a current software project as an example. What type of team do you think it should have? Evaluate your choice, enumerating pros and cons.
3. Conduct with your team the bonus allocation task described in this chapter. Based on the different reward allocation policies described in the task, identify factors that may lead your team members to cooperate. What are your conclusions with respect to the way your team works and cooperates?

REFERENCES AND ADDITIONAL RESOURCES

[Ambler03] Ambler, Scott, "Outsourcing Examined," *Software Development,* April 2003.

[Beck00] Beck, Kent, *Extreme Programming Explained,* Addison-Wesley, 2000.

[Brooks95] Brooks, P., Frederick, *Mythical Man-Month,* Second Edition, Addison-Wesley, 1995.

[Brown95] Brown, D. I., "Team-Based Reward Plans," *Team Performance Management* (January 1995): pp. 23–31.

[Duarte01] Duarte, Deborah L., and Snyder, Nancy Tennant, *Mastering Virtual Teams,* Second Edition, Jossey-Bass, 2001.

[Hayes03] Hayes, Frank, "Outsourcing Angst," *Computerworld* (March 17, 2003): p. 62.

[Hayes04] Hayes, Tim, "Outsourcing," *Pittsburgh TEQ* (2004), 10, 1, pp. 14–20.

[Hazzan03] Hazzan, Orit, "Computer Science Students' Conception of the Relationship between Reward (Grade) and Cooperation," *Proceedings of the Eighth Annual Conference on Innovation and Technology in Computer Science Education (ITiCSE 2003)*, Thessaloniki, Greece, 2003: pp. 178–182.

[Humphrey95] Humphrey, Watts, *A Discipline for Software Engineering,* Addison-Wesley, 1995.

[Humphrey00] Humphrey, Watts, *The Team Software Process*, Addison-Wesley, 2000.

[King03] King, Julia, "IT Global Itinerary," *Computerworld* (September 15, 2003): pp. 26–27.

[Mills83] Mills, Harlan D., *Software Productivity,* Little, Brown, 1983.

[Pressman03] Pressman, Roger, *Software Engineering: A Practitioner's Approach*, Fifth Edition, McGraw-Hill, 2003.

[Redmond01] Redmond, M. A., "A Computer Program to Aid Assignment of Student Project Groups," *The Proceedings of the 32nd SIGCSE Technical Symposium on Computer Science Education*, 2001: pp. 134–138.

Agile programming: *http://www.agilealliance.com/home*
Better estimates through function points: *http://www.ifpug.org/*
Project management: *http://www.pmi.org/info/default.asp*
Software Engineering Institute site: *http://www.sei.cmu.edu/*
The Capability Maturity Model for software: *http://www.sei.cmu.edu/cmm/cmm.html*

ENDNOTES

[1] Jim Tomayko once worked at a plant that combined manufacturing computers and engineering. A language lawyer worked there as well. When a program failed, he could read the hexadecimal inside the registers and tell you what was happening. He was *very* sharp.

4 Software as a Product

In This Chapter

- Introduction
- Objectives
- Study Questions
- Relevance for Software Engineering
- Software Requirements—Background
- Data Collection Tools
- Requirements Management
- Characteristics of Tools for Requirements Management
- Summary Questions

INTRODUCTION

This chapter examines the human aspect of software engineering from the perspective of the customers—the people who use software products. It focuses on how users and developers deal with defining software requirements in a way that fulfills customers' needs. Indeed, the process of defining customers' requirements is viewed here as a process in which both the customer and the developers participate. This approach was illustrated in previous chapters as well; for example, in Chapter 2, "Software Engineering Methods," in which eXtreme Programming (XP) is discussed.

Requirements are discussed in this chapter by looking at two topics: data collection tools and requirements management. The former deals with how to collect data in order to define customers' requirements; the latter deals with how, in practice, software developers manage the development of the defined requirements.

OBJECTIVES

- Readers will conceive of the customer as part of the human aspect of software development.
- Readers will increase their awareness with respect to the fact that requirements change during the process of software development.
- Readers will be familiar with data-gathering tools that may help them in formulating customers' requirements.
- Readers will be familiar with the concept of requirements management and tools that may support this process.

STUDY QUESTIONS

1. Why do we write software? What are the purposes of code writing?
2. What human beings' needs do software systems serve?
3. What software projects would you consider successes? Failures?
4. In your opinion, what percentage of software products succeeds? Fails?
5. According to [Mullet99], three quarters of all large software products delivered to the customer are failures. Sometimes they are not used at all; in other cases, they do not meet the customer's requirements. How would you explain this phenomenon? Where do the problems lie?
6. Based on Chapter 2, how do different software development methods attempt to develop software systems that fit customers' needs?

RELEVANCE FOR SOFTWARE ENGINEERING

If you ask experienced software developers for the five most serious problems of software engineering, it is reasonable to assume that one of the problems would be "requirements are changed." For example, Fairley and Willshire present a list of 10 problems and some antidotes for software projects. On that list, "changing needs" is the second item, and "requirements creep" is the seventh [Fairley and Willshire03]. Needless to say, a customer's conception of the developed application changes is independent of what software development method is used. "Requirements change" is a given fact in software development. Customers should not be blamed for changing their requirements. It is simply impossible to predict at the beginning of the software development process how the conception of the developed software system will be shaped.

When requirements are changed and the final product does not meet the customer's expectations, software developers may still claim that their development process is accomplished perfectly, in order to justify their creation. They can even be right. However, since software success is also measured according to how well it meets the customers' requirements, if customers change their mind during the course of software development, changes should be made in the software. If the software is not used eventually because it does not fit customer's needs, it does not matter whether it was developed adequately according to some established software development method. One conclusion that can be drawn from this fact is that when you consider what software development method to follow, you should also check whether it supports changes and updating of requirements that are introduced during the development process.

This chapter attempts to increase readers' awareness of the fact the requirements cannot be formulated formally and completely in advance. Some freedom for changing and updating should be provided. This chapter presents characteristics of tools that enable the management of these changes.

How to start the collection of the requirements in the first place? An answer to this question is presented in the first part of this chapter, which is dedicated to methods for data collection that can be used for defining software requirements.

SOFTWARE REQUIREMENTS—BACKGROUND

Software tools are developed for different reasons. For example, some people write software for their enjoyment; others write software to fulfill course requirements; yet others develop software systems to solve a problem they face. This chapter examines software systems that are developed for real customers in order to fulfill a customer's need. Relevant questions that should be asked at this stage are: How do customers define their needs? How can software developers help customers define the requirements?

In Chapter 2, we discussed how different software development methods deal with customer requirements. For example, the Unified Process describes use cases, which are what the user wants a computer system to do. The use cases usually represent the highest risks, and the requirements can be derived from them. Another approach to requirements definition is presented by eXtreme Programming (XP). XP advocates that in order to produce the software the customer needs, the customer should be on-site to give the developers ongoing feedback. In addition, XP outlines a very specific process (the Planning Game) for defining customer requirements (called "customer stories").

As hinted before, the specific development method one uses is of less importance as long as it ensures that customers get what they need. From this perspective, in the continuation of this chapter we look at how software requirements may be gathered.

TASKS

1. Usually, students are given a list of requirements and are asked to develop a software system that meets these requirements. This task may help you experience some of the problems involved in defining software requirements. For this purpose, you are asked to assume that you are a customer who needs a software system for Web-based surveys.

    First, determine the type of business you have. Based on this decision, define your requirements for the Web-based system. Write these requirements as a person who is not a software developer.

    After you finish listing the requirements, analyze them:

    - What kinds of requirements did you list (user oriented, technology oriented, performance oriented, others)?
    - Compare the list you defined with real Web-based survey tools. Can you use existing tools as a resource for gathering requirements?

    This exercise is important for at least two reasons. First, you may be a customer of software systems and you will have to define the requirements of the software systems you need. Second, as a software developer, when you have real customers, such an experience may help you see the situation from the customer's point of view.

2. Based on your experience in Task 1, explain why requirements change is so predominant in software engineering.
3. As previously mentioned, data indicates that the percentage of software tools that meet customers' needs is relatively low. Based on your experience in working on Task 1, explain this phenomenon.
4. Are there situations in which the software developers should participate in the gathering of information about the customer in order to help customers define their needs? If yes, what characterizes these situations? Why in these cases can't the software developers rely only on how customers define their requirements?

Discussion

This discussion is based on the requirements lists created in Task 1. During this discussion, each list can be checked, for example, according to its completeness and consistency. After several lists are reviewed, we recommend checking whether one system—that is, one characterized by combining several requirements lists—can be developed. In this way, we may get one product and the company may become a product-based company (in contrast to a company that has several versions of the same product, each version tailored for a specific customer).

If, during that discussion, requirements are changed, updated, altered, and so forth, discuss the reasons that lead to these changes. This is, in fact, how the process of defining requirements happens in everyday life and it is an excellent opportunity to reflect on that process (see Chapter 10, "Learning Processes in Software Engineering").

DATA COLLECTION TOOLS

The need to gather information about customers emerges in cases when the required system is complex, the customer does not know how to define the requirements, and knowledge about the best way the computerized system can serve the customer is required. There are several ways to gather information about customers in order to assist them in defining their requirements. The tools discussed in this chapter are interviews, observations, and questionnaires. We highlight how these tools are used for gathering qualitative data (to distinguish from quantitative data), since we are interested in people's conceptions, beliefs, and needs. In practice, for the development of real systems, we need to collect both qualitative and quantitative data.

To gain different perspectives and a wide as possible view of the customer's needs, in most of the cases more than one data collection tool is used. For example, in a questionnaire, we may ask about the customer's opinion; in an observation, we can look at how the customer behaves and analyze what this behavior implies for the developed software; and, in interviews, we may ask the interviewee to elaborate and explain what was written in the questionnaires and what was observed in the observation.

Interviews

There are different kinds of interviews: TV interviews, job interviews, ethnographic interviews, and others. Although they are all conducted for different aims, they all have one target: the interviewer wants to learn something about the interviewee that the interviewer doesn't already know. This "something" can be the interviewee's opinions, interpretations, and simple facts about work routines.

The aim of interviews conducted to gather information about a customer who asks for a software system is to learn about the features of the needed software, the anticipated user interaction with the system, and any other information that may improve the developers' understanding about the customer's needs related to the developed software. In general, when we come to interview future users of the application to be developed, we should try to understand as much as possible the interviewee's point of view.

Interviews are a very useful way to get information. Interviewees may present information relevant to their work, express their opinion and estimations with respect to different aspects of the application, and articulate their feelings about their work and the expected influence of the new system on their work. Of course, the information gathered during an interview may be inaccurate, but interviews supply us with first-hand information. Still, we should not make any decision about the requirements of the developed system based only on interviews. As mentioned previously, data should be collected using other tools as well.

In the following discussion, we assume that interviews are conducted as a preparation for the development of an application that will be used by many users of different types, all in one organization. This assumption may raise interesting problems that may not emerge in cases in which an application is developed for only one person. Here are several questions that we should ask ourselves when we face such a situation: Who will be interviewed in the organization? If we know whom to interview, in what order should we interview them? What types of questions should we ask each person, and how will the interview be managed? Should we decide on the interview's structure *a priori*, or should we act according to how the interview proceeds? These questions are addressed in the paragraphs that follow.

Of course, there is no need to interview all the employees of the organization. We should select the key people from any sector (group) that will use the system. It is important to select people from different groups and levels, as they may interpret various issues differently. Furthermore, sometimes it may be informative to interview people who are not part of the organization but are in contact with people in the organization. Their input may help in creating a system that also supports the organization's connections with other parties. The decision about who to interview can also be made based on interviewing key people and observing the routine of the organization.

After the group of interviewees is selected, we should determine what questions to ask them. Before the specific formulation of the questions, we should clarify what information we want to obtain from each sector. For example, we may be interested in issues such as the way the work is managed without the software that is going to be developed and the expected influence of the new system. After we clarify the subjects of the questions, we should determine what information we need with respect to each subject. For example, are we interested in facts or in the interviewee's opinion? With these ideas in mind, we can formulate the specific questions.

Let us focus on an example. Assume that your software house is asked to develop an online system for managing bids in the construction area. The system should support the bidding process, provide information about the products that are sold, and present any other information that may be useful for anyone who enters the Web site. The company that asks for this application is a big factory, that by offering the bid service, wants to become the main arbitrator in the area.

We first answer the question, "who should be interviewed?" Of course, the person who will operate the business aspect of the system should be interviewed. This person will check that the system works properly, that all the procedures work legally, and that the entire process goes smoothly. This person's perspective and input are very important. Another person who should be interviewed is an expert in what happens in this industry. This person may help us understand who will use the application, what type of information these people will look for, and so forth. Of course, other groups of people should be interviewed. For example, it will be useful to interview potential users of the Web site who are not part of the organization and those who will supply the products that will be purchased from this factory online via the Web site.

Now, we focus on the questions that will be posed to the different groups. As an example, we examine the questions that may be posed to the potential users of the Web site. We may want to learn about their current problems in getting the services they need, tools that may improve their work, types of information they find difficult to get, and their opinion with respect to the proposed tool. Each of these topics can be translated into a set of questions.

TASK

Translate the aforementioned topics to specific questions to be asked in an interview. For example, the subject "types of information they find difficult to get" can be referred in an interview by questions such as:

- On a scale of 1 to 5, indicate how difficult it is for you to find information related to your business.
- In your opinion, what type of information is crucial for decision-making processes in your field?

The questions presented in the preceding task indicate that different types of questions can be presented; for example, open questions (which usually start with phrases like "In your opinion ... ," "Can you elaborate ... ," "Can you describe ...") and closed questions (such as "On a scale of ..."). Indeed, in an interview it makes more sense to present open questions, as they invite the interviewee to express opinions that are more difficult to be expressed in a written questionnaire. Furthermore, when interviewees present their opinions with respect to a specific issue, we can ask additional questions when we want to deepen our understanding with respect to that topic. This is impossible to do in a questionnaire whose questions are determined in advance and cannot be changed after its distribution. We elaborate more on types of questions in the discussion about questionnaires.

There are opinions that suggest presenting different types of questions to people at different levels in the organization. Accordingly, these opinions suggest presenting to people in high-level positions questions about the organization's targets, problems, and budget constraints and presenting to employees in lower levels questions about the daily work routine. We would like to propose a different perspective. We believe that it is important to hear what people from different levels in the organization think about topics to which people from other levels in the organizations are more closely related. Thus, for example, we can learn a lot from listening to the perception of a junior employee about the organization's targets and main problems. Similarly, when we pose a senior manager a question about the main routines of the organization, we may learn how the manager conceives of the lower-level employees' work.

An interview can be managed in different ways: structured, semi-structured, or not structured. The difference between the three types of interviews lies in the level of freedom we have to detach from the sequence of questions that we prepared prior to the interview. In a structured interview, we follow the questions in the order in which they were prepared; in a not-structured interview we allow ourselves to present new questions, to change the order of the questions, to ask for clarifications, and so forth, according to the interview progress. In a semi-structured interview, this freedom is given but is limited. Of course, the questions prepared for the interview influence the flow and the type of interview. The more closed the questions are, the more structured interview we get.

After deciding whom to interview and what to ask each interviewee, we should decide on the order of the interviews. We can simply decide that we conduct the interviews when the interviewees are available. However, we may get additional information from conducting the interviews in a specific order. As there is no unique answer to the question of order, we suggest intertwining interviews from several sectors and levels. That is, not to interview all the managers first and then all the lower-level employees, but rather, in the sequence of interviews, to move between levels of seniority.

TASK
What benefits might we gain from intertwining interviews of people from different levels of seniority?

Now that we have decided whom to interview, what to ask each of them, and in what order to conduct the interviews, we should address the interviewer-interviewee relationship. In fact, this is one of the crucial factors that may determine the success of the interview. It is very important to set a comfortable atmosphere in the interview, to eliminate as much as possible status differences, to listen very carefully to what is said, and to respond to new directions that the interviewee raises during the interview. In addition, it is very important not to guide the interviewee to answer a specific answer that we may want to hear (remember, there are no correct or incorrect opinions!).

The last issue to consider with respect to interviews is their recording. The recording may help us in later stages when we want to analyze what was said and what we heard. Ethical conventions tell us to inform the interviewees that they are being recorded and to explain the target of the recording.

Activity
Think about three customers of different sizes (e.g., a local book store, an insurance company, and an international bank). Assume that you are asked to develop a software system for managing each. You decide to start learning about these organizations by interviews. For each company, work on the following tasks:

- What groups in the organizations will you interview?
- What questions will you ask each group of interviewees?
- How will you establish a good atmosphere in the interviews?

After finishing working on these tasks, reflect: what guidelines helped you answer the questions with respect to each of the three organizations? Are these guidelines different?

Questionnaires

All of us fill in questionnaires in different situations. For example, we may be asked to fill in a questionnaire about a product we purchased; in college, we fill in teaching evaluation surveys. In all cases, those who answer the questionnaires are asked to express their point of view and to supply relevant facts. Similarly, questionnaires may be filled in as part of the process of defining the requirements of a software system. Naturally, in the latter case the focus is placed on the software system to be developed.

Questionnaires help us get, in relatively short time, a lot of data from many people in different places. To gain different types of data we use different types of questions. Following the brief discussion about the kinds of questions we presented in the section about interviews, we refer here to two categorizations of questions: open/closed questions and fact/opinion questions. Table 4.1 presents examples for each type of question.

To get some feedback about how the questions posed in the questionnaire are understood, it is recommended to distribute it first to a small number of people, to see how they answer the questions and to check whether a particular formulation should be improved. When the real questionnaires are returned it is important to remember that if the percentage of the responses is low, we may get diversion in the results, as those who answered the questionnaires may have a specific characteristic that encouraged them to answer and return the questionnaire. This is another reason why different tools for data gathering should be used.

TABLE 4.1 Examples of Questions

	Closed Question	**Open Question**
Fact Question	On an average day, how many hours do you dedicate to e-mail? a. less than 30 minutes b. more than 30 minutes, less than an hour c. between 1 and 5 hours d. more than 5 hours	What are the three most used features of the application? Describe how you use each.
Opinion Question	In your opinion, what percentage of the employees would like to work with the new system? a. less than 25% b. between 25% and 75% c. more than 75%	In your opinion, how will the new application influence the workflow of each of the main players in your organization?

TASK

For one of the companies on which you worked at the end of the section about interviews, compose two questionnaires. One questionnaire uses closed questions; one questionnaire consists of open questions. Answer the following questions:
- What kind of information will you get from each questionnaire?
- What helpful information does each questionnaire provide?
- What does each questionnaire lack?
- How can each questionnaire be improved?

Observation

This method for data collection is borrowed from the ethnographic research arena. Its purpose is to identify people's behavioral patterns in different situations. In the case of software development, we are interested in how people may work with the software to be developed. We should be aware of the fact that there are cases in which people use (software) tools in different ways than those that were planned originally.

Observation is appropriate to be used when requirements are observable; for example, in service work. In general, observations may teach us about relationships between workers and between workers and customers, emotional responses, and so forth. Sometimes we need to observe a process, and we have to move between locations; sometimes we have to stay in one place in order to learn about a specific subject. As with questionnaires, we may use a structured observation (in which the objects to be observed are determined in advance) or unstructured observation (in which we observe what seems to be interesting and relevant even if it was not determined in advance when we planned the observation). In all cases, we should remember that those who are observed may change their behavior when we look at their activities. This fact should be taken into consideration when the observation is analyzed.

TASKS

1. The Hawthorne effect is a research phenomenon in which research results are distorted in the direction expected but not for the reason expected. How is the Hawthorne effect relevant to data collection by observations? How would you cope with its implications? Further information about the Hawthorne effect can be found on the Web.
2. Continue working on gathering information on the company for which you developed the interview and the questionnaires. Select a topic that you would like to observe in that company. Answer the following questions:
   - Why should this topic be observed?
   - What type of information might you get from observing this subject?

- What type of observation (structured or unstructured) would you perform in this case? Why?

REQUIREMENTS MANAGEMENT

The first part of this chapter, which deals with the gathering of information, can be considered part of the management of requirements. Indeed, no matter at what stage a requirement is introduced, it should be managed. In practice, the more the software world is developed, the more it is accepted that the way in which requirements are managed is a critical factor of software success. Efforts made in this direction come to improve software quality at least from the customer's perspective.

The management of requirements can be discussed from a technical point of view (that is, what tools can be used for this purpose and how to use them) and from the human perspective (that is, what needs this activity serves in the process of software development). Naturally, in what follows, the emphasis is placed on the human aspect and not on technical features of any particular tool.

TASKS

1. What are the targets of requirements management tools?
2. What should characterize these tools?
3. How might such tools influence the process of software development?
4. Assume that you are going to establish a company that offers a tool for requirements management. Identify your customers. What would you offer them?
5. In your opinion, can requirements be managed by traditional project management tools? If yes, how? If not, in what sense does a software project differ from other kinds of projects, and, consequently, requires different tools for managing its requirements?

In what follows we present a summary of basic features that computational tools for requirements management should have, emphasizing their importance from the human perspective. We do not claim that the list of features is complete. However, it contains main features that support this complex process. The list is composed based on the analysis of several tools for requirements management (listed at the end of this chapter).

CHARACTERISTICS OF TOOLS FOR REQUIREMENTS MANAGEMENT

When listing features of tools for requirements management, we assume that the tool is computerized, and that the tool is "easy to use." Clearly, if a tool is not easy to use, it will not be used no matter what important features it offers.

I. Enable basic operations that are performed on requirements: Such tools should provide the option to define and list requirements in an *incremental* process during the course of gathering the requirements and, after that, during the entire process of software development. At any stage, when requirements are listed, such tools should support the performance of changes in the requirements that result from the customer's improved understanding of the needed software. As declared previously, customers cannot define *a priori* all the requirements. Thus, these tools should support the ability to change, update, and refine requirements.

If the requirements are changed during the development process, such tools should support the tracing of the requirements development together with an explanation of why they are changed.

Here are additional basic operations on requirements that such tools should offer:

- **Allocation to developers:** Who is in charge of the development of each requirement?
- **Verification:** What (automatic) test shows that a feature is completed?
- **Tracking the status of any requirement:** Is a requirement in development? Is it tested? Is it completed?
- **Means for notification when deadlines, changes, or status modifications happen:** Was a specific deadline changed? Why? What effects does this change have on other deadlines?

II. Show interconnections between requirements: It is important that tools for requirements management show links between requirements (how they are connected to each other) so that changes are cross-notified and relevant changes do not slip through. To keep the customer's priorities intact, such links should include prioritization; that is, which requirement is more important. A graphical presentation of that hierarchy may help the developers identify which requirements are impacted by what changes.

III. Support teamwork and communication: The way in which requirements are managed and developed should be transparent to all the team members. It is clear that the better the communication between team members, the better they can communicate the requirements. This, in turn, leads to the development of a software tool that meets the customer's needs in a better way. When such a tool is accessed by all team members, it improves the development process, and can be used for improving communication between the development team and the customer and between different functions in the company. Thus, for example, senior management may gain a clearer picture of what is developed; customer service representatives get follow-up notification about which version includes their requirements; and communication between the development and the marketing departments is improved. Consequently, situations in which the marketing people promise the customers features that the development department is not aware of may be avoided. To gain this cross-organization access to the requirements management tools, such tools should be collaborative. Naturally, Web-based tools match this purpose.

IV. Integration with other development tools: To improve the information flow and interactivity between all the development activities, a requirements management tool may be integrated with the configuration management tool that the team uses. This feature is important only if it does not increase the complexity of the configuration management tool and is easy to use. Similarly, if it does not add to the usage complexity, it may also be integrated with documents and spreadsheets related to the project.

TASKS

1. The preceding list of characteristics of requirements management tools is a first step in the analysis of the requirements of a tool that manages requirements. These requirements may be changed during the actual development of the tool. Assume that you intend to develop such a tool using an object-oriented approach. What classes and methods will your system consist of?
2. Assume that in your organization, requirements are not managed properly. You decide to initiate the introduction of a requirements management tool. How would you start this process? By convincing the top management? By convincing the developers? What advantages does each approach have? What pitfalls does each approach have? What problems might you face in this process?
3. Discuss connections between requirements management and teamwork (see Chapter 3).

4. Discuss connections between requirements management and learning processes in software engineering (see Chapter 10).

DISCUSSION

As it turns out, although coding is the activity that produces software systems, it is not the only main issue of software development processes. Other activities conducted during the development process also have a significant influence on software project success. Two of these activities are the gathering and management of requirements. As mentioned previously, these issues are so important that some argue that no matter how a software system is developed, if the processes of defining the requirements and their management are done properly, the rest of the development process continues smoothly. How much effort is invested in the management of requirements is an organizational decision that is partially based on the organizational culture.

1. Connect the topics discussed in this chapter to the Code of Ethics of Software Engineering (see Chapter 5, "Code of Ethics of Software Engineering").
2. In some organizations, knowledge is conceived as power. How might this fact influence the process of requirements management?

SUMMARY QUESTIONS

1. Assume that you have to start gathering information related to a new software system for an organization that currently does not use any computational system. How would you start collecting the data? Why?
2. Analyze three software tools that you use in your everyday work activities. What are their main characteristics? What customers' requirements do they reflect? Can you suggest improvements for these tools?

FOR FURTHER REVIEW

1. Visit a company (a software house or any other company). Observe how people communicate and behave in that company. Identify a situation in the company workflow that can be improved by a computational tool. Create a requirements list for this tool. Interview different people in the organization about this list of requirements. Reflect: Do their impressions fit

yours? Do they suggest improvements? How would you improve the requirements list based on these interviews?
2. Companies that produce products in the same domain tend to be constantly working with the same or similar requirements. Many now are choosing product lines, as these are more productive and more important in the long term [Clements01]. Product lines are most effective when there is a large amount of reuse. A company or team figures out which parts of the software can be an element of another product, and this element is made in such a way that it can easily be part of a new product. Explore additional benefits of reuse in software development. Focus on the human aspects of these benefits.

REFERENCES

[Clements01] Clements, Paul, and Northrop, Linda, *Software Product Lines: Practices and Patterns,* Addison Wesley: Boston, 2001.

[Fairley and Willshire03] Fairley, Richard E., and Willshire, Mary Jane, "Why the Vasa Sank: 10 Problems and Some Antidotes for Software Projects," *IEEE Software* (April–May 2003): pp. 18–25.

[Mullet99] Mullet, Dianna, "The Software Crisis," Benchmarks Online—a Monthly Publication of Academic Computing Services—a Division of the University of North Texas Computing Center 2(7), July 1999, *www.unt.edu/benchmarks/archives/1999/july99/crisis.htm*.

Examples of Requirements Management Tools Reviewed

- **RMTrk:** *www.rmtrak.com/*
- **Rational® RequisitePro®:** *www.rational.com/products/reqpro/index.jsp?SMSESSION=NO*
- **Goda Software:** *www.analysttool.com/*
- **SpeeDEV:** *www.speedev.com/speedProcess.html*
- **Telelogic:** *www.software-systems-development.com/*
- **Intrinsyx Technologies:** *www.intrinsyx.com/technologies/services/requirements.htm*

Part II
The World of Software Engineering

Chapter 5, Code of Ethics of Software Engineering
Chapter 6, International Perspective on Software Engineering
Chapter 7, Different Perspectives on Software Engineering
Chapter 8, The History of Software Engineering

5 Code of Ethics of Software Engineering

In This Chapter

- Introduction
- Objectives
- Study Questions
- Relevance for Software Engineering
- Codes of Ethics
- The Code of Ethics of Software Engineering
- Scanning the Code of Ethics of Software Engineering
- Discussion
- Summary Questions

INTRODUCTION

In this chapter, we discuss the notion of ethics in general and review the Code of Ethics of Software Engineering[1] in particular. These aims are achieved by discussing the main ideas and the essence of the Code of Ethics of Software Engineering and by analyzing scenarios taken from the software engineering world. This chapter is largely based on case studies that the readers are asked to discuss. The idea is to guide the reader in the formulation of personal ethical behavior. This is a personal process that each practitioner should go through individually. In this sense, this chapter is different from the other chapters of the book since it raises questions and philosophical dilemmas more than it provides answers.

OBJECTIVES

- Readers will become familiar with the concept of code of ethics in general and the Code of Ethics of Software Engineering in particular.
- Readers will identify situations in which ethical considerations (in addition to technical considerations) should be intertwined in software development processes.
- Readers will improve their ability to predict situations in software development processes that may harm society from an ethical point of view and to suggest ways to avoid such situations.
- Readers will conceive the Code of Ethics of Software Engineering as a tool that can be used for both the identification of ethical dilemmas and solving them.

STUDY QUESTIONS

1. Define basic rules that software engineers should follow when they develop any piece of code. Refer to "to do" rules and "not to do" rules.
2. In your opinion, in what ways might software developers influence our world? Refer to social, environmental, and other types of influences.
3. In your opinion, in what ways might software developers influence a software user's life? How is it possible to protect users from harmful influences?
4. In your opinion, in what ways can software engineers harm their coworkers? How can such harmful behavior be avoided?
5. This question addresses the idea of ethics. You are asked to answer it according to your familiarity with the concept.

    What is ethics? What professions have a code of ethics? If you are familiar with other professions that have a code of ethics, on what fundamental rules are their codes of ethics based? What types of problems do their codes of ethics address?
6. Identify five situations in which ethical issues should be considered in software development.

    In your opinion and based on your familiarity with the notion of ethics and the situations you just described, does the community of software engineering need a code of ethics? If "yes," explain why. On what principles should it be based? What topics should it address? If "no," explain your opinion.
7. What special ethical problems might the Internet raise? In what sense are these issues different from other ethical situations related to software engineering?

RELEVANCE FOR SOFTWARE ENGINEERING

There is no doubt that software developers have a significant impact on our environment, community, and culture. Software tools influence how people live, communicate, and (sometimes) behave. This chapter emphasizes that these influences should be considered when one develops any software system. At the same time, it is important to determine the ethical considerations one should consider in the process of software development. The Code of Ethics of Software Engineering comes to address at least partially some of these issues and to guide software developers when sensitive and ambiguous issues arise.

This chapter attempts to highlight several guidelines, recommended within the framework of the Code of Ethics of Software Engineering, that one should take into consideration in the process of software development. These guidelines address both the development environment and the potential influence outside the development environment of what is developed. Although a code of ethics can be addressed on different levels (for example, individual, team, organization, and country), we limit the discussion to the individual level. We hope that the detailed discussion with respect to the individual level will establish a basis for the discussion about the application of the Code of Ethics of Software Engineering to the higher levels.

CODES OF ETHICS

Some communities of practice have a well-known code of ethics (The Code of Medical Ethics, for example). The role of such codes is to guide professionals on how to behave in situations when the right action may be unclear. The need for such a code arises from the fact that any profession generates situations that can neither be predicted nor addressed uniformly by all members of the community. At this stage, you may ask yourself, "Does the software engineering community need a code of ethics?" If yes, what situations are appropriate to be addressed by this code of ethics? What situations should be addressed by the law? This section starts by examining the concept of ethics in general. Then, we focus on the Code of Ethics of Software Engineering.

Ethics is part of the discipline of philosophy. *The New Shorter Oxford English Dictionary* defines ethics as "the science (or set) of moral principles; the branch of knowledge that deals with the principles of human duty or the logic of moral discourse." *Webster's Collegiate Dictionary* adds that ethics is "the discipline dealing with what is good and bad and with moral duty and obligation." *Webster's Dictionary* specifies that ethics consists of "the principles of conduct governing an individual or group."

Although there is a distinction between "the science of ethics" and "applied ethics" (the application of the theories of ethics to the real world), it is not our intention here to go into detail with respect to this distinction. Rather, we focus on the application of the Code of Ethics of Software Engineering to real-life situations that software developers may face. In what follows, we focus on professional ethics and adopt the practical definition for the concept. According to this definition, ethics indicates how professionals should behave in situations that arise in their profession. This behavior should be based on moral norms and the professionals' ability to discern what is right and what is wrong.

Discussion

1. Suggest and discuss situations in which a code of ethics may disturb professionals' work. What should be done in such cases?
2. Suggest and discuss situations in which a code of ethics may be abused. What should be done in such cases?

Using practical terminology, we might say that ethics guides daily professional behavior by values that are less formulated than a "to do" or "not to do" list of rules. In other words, we might say that a professional code of ethics, like the Code of Ethics of Software Engineering, reflects the moral norms of the profession. However, it is important to note that ethics is *not* based on rules that are formulated in terms of punishment.

In what follows, we examine briefly what a profession is, address the basics on which ethical norms rely, and understand the place of the code of ethics in any profession. This discussion is based on Asa Kasher's analysis of the conception of being a professional [Kasher02].

The proposed conception of being a professional consists of five basic layers:

- **Systematic knowledge** of the relevant area.
- **Systematic proficiency in solving problems** within the relevant area: a toolbox or problem-solving skills within the profession domain.
- **Constant improvement of that knowledge and proficiency,** meaning the duty to keep up-to-date. Systematic knowledge and the skills are dated.
- **Local understanding of professional claims and methods** in a way that enhances problem-solving ability. This layer refers to the ability to argue and reason about the professional activities and the ability to answer "why" questions. The importance of this understanding is reflected in situations in which the toolbox lacks an appropriate solution.

- **Global understanding of professional activity,** understanding of the essence of the profession. Accordingly, ethics sets superior behavior standards. The main resources for any ethics are the professional dignity, the professional community, and the requirements of the surrounding social environments (democratic values, for example).

TASK

For each of the aforementioned five layers, give an example with respect to software engineering.

When appropriate, organizations adopt their own ethics common to all the professions within the organization, and completes the code of ethics of each profession. The following discussion focuses on software engineering.

THE CODE OF ETHICS OF SOFTWARE ENGINEERING

Generally speaking, the importance of professional ethics for software engineering stems from the vast influence of computers in general and software in particular on life on Earth. On the one hand, software engineers may cause a lot of harm; on the other hand, software engineers may contribute a lot to the quality of life. Thus, software engineers should be aware of what is legitimate and what is forbidden, when they have the freedom to make critical decisions and when they do not, and so forth. The importance of the Code of Ethics of Software Engineering seemed obvious during the 1990s, when many startups focused on their financial success, sometimes without considering what values they deliver and according to what values they work.

TASK: SCENARIO ANALYSIS

In what follows, seven cases related to software engineering are presented. With respect to each scenario, express your opinion about the described behavior and describe how you would behave in such a case. Then, according to your decision, formulate one or more ethical norms that, in your opinion, should be included in the Code of Ethics of Software Engineering. These norms should guide software developers in similar situations.

Scenario One: Not Telling the Entire Truth

A programmer was asked to make a change to a software application used by an international bank. She performed all the needed tests. After all the tests passed, she recalls that one more test is required. This test does not pass. Since she does not have the time for debugging, she submits her work and states that all the tests passed successfully.

Scenario Two: Using a Software Tool Not Paid For

A freelance programmer is asked to develop a software project that requires using an expensive IDE (Integrated Development Environment) for two months. His friend, who works in a company that owns the IDE, suggests that he should borrow the IDE for the two-month development period and uninstall it when he completes his project. The programmer accepts the offer.

Scenario Three: Keeping Information from Management

A programmer who works in a software house hears from a friend that her team leader tells several competitive companies secret information about the next planned version she works on. Since the programmer predicts that she will not be able to go on working for the company if she tells management about the team leader's behavior, she decides to continue working as if nothing has happened.

Scenario Four: Causing Environmental Damage

A programmer is asked to work on a project that causes severe damage to the environment. This project enables his software company to survive. He tells his supervisor about his worries and is informed that part of the money the company will earn for this project will be donated to an environmental fund. He agrees to work on the project.

Scenario Five: Not Sharing Information with Teammates

A programmer is told by her team leader that if she is better than the others on the team, she will be promoted. She decides that she can be promoted only if she does not share essential information with her teammates and behaves accordingly.

Scenario Six: Re-Downloading Limited-Time Freeware

A software engineer needs particular software for completing his job on time. A demo version can be downloaded from the Web for a 14-day examination. The engineer keeps downloading the demo version every two weeks and does not purchase it.

Scenario Seven: Using Company Resources

A software engineer takes advantage of the free Internet access that her company offers and constructs a Web site that enables free downloading of musical files without paying for them.

Before we move on to review the Code of Ethics of Software Engineering it is important to note that in ethical discussions, there is no unique solution, and in most of the cases, the decision how to behave depends on one's values. This is the essence of ethics, and this characteristic of ethics distinguishes it from the law. This nature of ethics does not allow us, for example, to present strict and definite solutions and modes of behavior when ethical considerations are introduced. In general, each person should determine what his or her ethical values are. As it turns out, in most cases, people eventually work in an environment that reflects their ethical values and standards. For example, an ethical company hires ethical employees and has business relations with ethical partners.

In this spirit, the reader should not expect to find solutions to all the presented problems. Rather, the following discussion aims to help one formulate one's ethical values and to increase reader understanding on how ethical considerations may support software development work.

SCANNING THE CODE OF ETHICS OF SOFTWARE ENGINEERING

The Code of Ethics of Software Engineering appears in two versions: a short version and a full version. The short version highlights the main topics addressed in the code of ethics: public, client and employer, product, judgment, management, profession, colleagues, and self. The full version fills in details and adds specific rules that outline a particular mode of behavior. We first present the short version (see Table 5.1).[2]

TASK

Read the short version of the Code of Ethics of Software Engineering. For each of its sections, suggest situations in software development that this particular section might address. In other words, for each section of the code of ethics suggest situations for which it may be helpful in guiding software developers' behavior.

Before moving to the full version of the Code of Ethics of Software Engineering, it is important to note that the first principle in the previous version of the Code of Ethics of Software Engineering was the product, while in the updated version the first principle is the public. This change is more than a semantic update. It partially reflects the shift in the emphasis of what software engineering is. It is not sufficient anymore to produce qualified code; the development process should also stand in line with social values and the public interest.

The full version of the Code of Ethics of Software Engineering also strengthens the feeling that such a code should reflect the fact that the software world keeps changing. As we shall see, the details that are added to the full version of the Code of Ethics of Software Engineering are formulated in a way that keeps them relevant when changes and revolutions happen in the software world.

TABLE 5.1 Short Version—Software Engineering Code of Ethics and Professional Practice

SOFTWARE ENGINEERING CODE OF ETHICS AND PROFESSIONAL PRACTICE

(Version 5.2) as recommended by the

IEEE-CS/ACM Joint Task Force on Software Engineering Ethics and Professional Practices

Short Version

PREAMBLE

The short version of the code summarizes aspirations at a high level of abstraction. The clauses that are included in the full version give examples and details of how these aspirations change the way we act as software engineering professionals. Without the aspirations, the details can become legalistic and tedious; without the details, the aspirations can become high sounding but empty; together, the aspirations and the details form a cohesive code.

Software engineers shall commit themselves to making the analysis, specification, design, development, testing and maintenance of software a beneficial and respected profession. In accordance with their commitment to the health, safety and welfare of the public, software engineers shall adhere to the following Eight Principles:

1 PUBLIC - Software engineers shall act consistently with the public interest.

2 CLIENT AND EMPLOYER - Software engineers shall act in a manner that is in the best interests of their client and employer, consistent with the public interest.

3 PRODUCT - Software engineers shall ensure that their products and related modifications meet the highest professional standards possible.

4 JUDGMENT - Software engineers shall maintain integrity and independence in their professional judgment.

5 MANAGEMENT - Software engineering managers and leaders shall subscribe to and promote an ethical approach to the management of software development and maintenance.

6 PROFESSION - Software engineers shall advance the integrity and reputation of the profession consistent with the public interest.

7 COLLEAGUES - Software engineers shall be fair to and supportive of their colleagues.

8 SELF - Software engineers shall participate in lifelong learning regarding the practice of their profession and shall promote an ethical approach to the practice of the profession.

©1999 by the Institute of Electrical and Electronics Engineers, Inc. and the Association for Computing Machinery, Inc.

This Code may be published without permission as long as it is not changed in any way and it carries the copyright notice.

Tables 5.2 through 5.11 present the full version of the code of ethics. Table 5.2 presents the introduction and the rationale of the code; Tables 5.3 through 5.10 present the sections of the code according to the eight principles; and Table 5.11 presents the code signature. The following activities address the code as a whole and refer mainly to situations in which one might or might not use the code of ethics.

Activities

Read the full version of the Code of Ethics of Software Engineering (Tables 5.2 through 5.11).

1. Describe situations in which the code of ethics may be helpful and situations in which it can raise conflicts. What types of conflicts might it raise?
2. Present a situation in the past in which you did not know how to behave. How could you have used the code of ethics in that case?
3. How much freedom does the code of ethics leave to software developers? With respect to what topics is this freedom given?
4. What sections are difficult to use and to work according to? What sections are easy to implement?
5. What sections/items would you omit? add? change?

TABLE 5.2 Software Engineering Code of Ethics and Professional Practice Full Version, Preamble

SOFTWARE ENGINEERING CODE OF ETHICS AND PROFESSIONAL PRACTICE

IEEE-CS/ACM Joint Task Force on Software Engineering Ethics and Professional Practices

Full Version

PREAMBLE

Computers have a central and growing role in commerce, industry, government, medicine, education, entertainment and society at large. Software engineers are those who contribute by direct participation or by teaching, to the analysis, specification, design, development, certification, maintenance and testing of software systems. Because of their roles in developing software systems, software engineers have significant opportunities to do good or cause harm, to enable others to do good or cause harm, or to influence others to do good or cause harm. To ensure, as much as possible, that their efforts will be used for good, software engineers must commit themselves to making software engineering a beneficial and respected profession. In accordance with that commitment, software engineers shall adhere to the following Code of Ethics and Professional Practice.

TABLE 5.2 (continued)

> The Code contains eight Principles related to the behavior of and decisions made by professional software engineers, including practitioners, educators, managers, supervisors and policy makers, as well as trainees and students of the profession. The Principles identify the ethically responsible relationships in which individuals, groups, and organizations participate and the primary obligations within these relationships. The Clauses of each Principle are illustrations of some of the obligations included in these relationships. These obligations are founded in the software engineer's humanity, in special care owed to people affected by the work of software engineers, and in the unique elements of the practice of software engineering. The Code prescribes these as obligations of anyone claiming to be or aspiring to be a software engineer.
>
> It is not intended that the individual parts of the Code be used in isolation to justify errors of omission or commission. The list of Principles and Clauses is not exhaustive. The Clauses should not be read as separating the acceptable from the unacceptable in professional conduct in all practical situations. The Code is not a simple ethical algorithm that generates ethical decisions. In some situations, standards may be in tension with each other or with standards from other sources. These situations require the software engineer to use ethical judgment to act in a manner which is most consistent with the spirit of the Code of Ethics and Professional Practice, given the circumstances.
>
> Ethical tensions can best be addressed by thoughtful consideration of fundamental principles, rather than blind reliance on detailed regulations. These Principles should influence software engineers to consider broadly who is affected by their work; to examine if they and their colleagues are treating other human beings with due respect; to consider how the public, if reasonably well informed, would view their decisions; to analyze how the least empowered will be affected by their decisions; and to consider whether their acts would be judged worthy of the ideal professional working as a software engineer. In all these judgments concern for the health, safety and welfare of the public is primary; that is, the "Public Interest" is central to this Code.
>
> The dynamic and demanding context of software engineering requires a code that is adaptable and relevant to new situations as they occur. However, even in this generality, the Code provides support for software engineers and managers of software engineers who need to take positive action in a specific case by documenting the ethical stance of the profession. The Code provides an ethical foundation to which individuals within teams and the team as a whole can appeal. The Code helps to define those actions that are ethically improper to request of a software engineer or teams of software engineers.
>
> The Code is not simply for adjudicating the nature of questionable acts; it also has an important educational function. As this Code expresses the consensus of the profession on ethical issues, it is a means to educate both the public and aspiring professionals about the ethical obligations of all software engineers.

Tables 5.3 through 5.10 present the code of ethics by principles. In between the tables, we pose questions that refer to the principle. In most questions, the reader is given a general description of situations that may occur in software development and is asked to fill in details. The idea is to show that the details may sometimes determine what decision should be made, and that small changes in the details may change our conception of the entire situation.

QUESTIONS—PRINCIPLE 1: PUBLIC

1. Suggest a situation in software development in which the members of a software development team have to make the decision whether to report to the authorities about a bug in a specific software tool that they developed. What does the code of ethics say in such cases? How would you behave in such a case?

TABLE 5.3 Software Engineering Code of Ethics and Professional Practice—Principle 1, Public

> **Principle 1 PUBLIC** Software engineers shall act consistently with the public interest. In particular, software engineers shall, as appropriate:
>
> Accept full responsibility for their own work.
>
> Moderate the interests of the software engineer, the employer, the client and the users with the public good.
>
> Approve software only if they have a well-founded belief that it is safe, meets specifications, passes appropriate tests, and does not diminish quality of life, diminish privacy or harm the environment. The ultimate effect of the work should be to the public good.
>
> Disclose to appropriate persons or authorities any actual or potential danger to the user, the public, or the environment, that they reasonably believe to be associated with software or related documents.
>
> Cooperate in efforts to address matters of grave public concern caused by software, its installation, maintenance, support or documentation.
>
> Be fair and avoid deception in all statements, particularly public ones, concerning software or related documents, methods and tools.
>
> Consider issues of physical disabilities, allocation of resources, economic disadvantage and other factors that can diminish access to the benefits of software.
>
> Be encouraged to volunteer professional skills to good causes and to contribute to public education concerning the discipline.

2. Suggest a situation in software development in which a software developer should address issues of physical (or other) disabilities, but an unrelated issue interferes with the developer doing so. What does the code of ethics say in such cases? How would you behave in such a case?
3. Your software house is asked to develop a software system for a company that causes air pollution but contributes a lot to the educational system in the area. The software system aims at increasing the production of the company, which might increase the air pollution. The support the company provides to the educational system is significant and may be increased when the production grows. Without this support, the education system would suffer significantly. How would you react if you were asked to take an active role in the development process?
4. Your friend works for a software house that specializes in the development of computer games. Recently, several publications have indicated that these games influence some children negatively. These games are the main product of your friend's company and, without them, the company may not be able to survive. The company's management is aware of these publications and gathers all the employees to discuss the future of the company. Assume you participate in the meeting and answer the following questions:
   - Suggest different opinions that might be expressed in the meeting. What ethical consideration does each opinion represent?
   - What conflicts of interest are presented in this case?
   - What is your opinion with respect to this case?
   - How would you behave in such a case?

QUESTIONS—PRINCIPLE 2: CLIENT AND EMPLOYER

1. Suggest a situation in which a software developer has to make the decision whether to report to clients that the software project ordered by them is likely to fail, when she knows that such a report may cause irreversible financial loss to her company. What does the code of ethics say in such cases? How would you behave in such a case?
2. A talented software engineer works for a software house and is offered a promotion if he helps the company find out some confidential information about the main client of the company. He struggles between his desire to be promoted and his high ethical standards. What ethical issues are raised in this case? How does the code of ethics address these issues? How would you behave in such a case? Under what circumstances, if any, can this behavior be considered ethical?

TABLE 5.4 Software Engineering Code of Ethics and Professional Practice—
Principle 2, Client and Employer

> **Principle 2 CLIENT AND EMPLOYER** Software engineers shall act in a manner that is in the best interests of their client and employer, consistent with the public interest. In particular, software engineers shall, as appropriate:
>
> Provide service in their areas of competence, being honest and forthright about any limitations of their experience and education.
>
> Not knowingly use software that is obtained or retained either illegally or unethically.
>
> Use the property of a client or employer only in ways properly authorized, and with the client's or employer's knowledge and consent.
>
> Ensure that any document upon which they rely has been approved, when required, by someone authorized to approve it.
>
> Keep private any confidential information gained in their professional work, where such confidentiality is consistent with the public interest and consistent with the law.
>
> Identify, document, collect evidence and report to the client or the employer promptly if, in their opinion, a project is likely to fail, to prove too expensive, to violate intellectual property law, or otherwise to be problematic.
>
> Identify, document, and report significant issues of social concern, of which they are aware, in software or related documents, to the employer or the client.
>
> Accept no outside work detrimental to the work they perform for their primary employer.
>
> Promote no interest adverse to their employer or client, unless a higher ethical concern is being compromised; in that case, inform the employer or another appropriate authority of the ethical concern.

QUESTIONS—PRINCIPLE 3: PRODUCT

1. A software developer believes that the development methodology adopted for the project he works on is inappropriate for that project. At the same time, he knows that the time needed for changing the methodology may postpone the delivery time to the customer. What does the code of ethics say in such cases? How would you behave in such a case?
2. A software developer feels she is not qualified to manage the project she has been assigned to lead, but she knows that if she admits it, she may be fired. What does the code of ethics say in such cases? How would you behave in such a case?

TABLE 5.5 Software Engineering Code of Ethics and Professional Practice—Principle 3, Product

Principle 3 PRODUCT Software engineers shall ensure that their products and related modifications meet the highest professional standards possible. In particular, software engineers shall, as appropriate:

Strive for high quality, acceptable cost, and a reasonable schedule, ensuring significant tradeoffs are clear to and accepted by the employer and the client, and are available for consideration by the user and the public.

Ensure proper and achievable goals and objectives for any project on which they work or propose.

Identify, define and address ethical, economic, cultural, legal and environmental issues related to work projects.

Ensure that they are qualified for any project on which they work or propose to work, by an appropriate combination of education, training, and experience.

Ensure that an appropriate method is used for any project on which they work or propose to work.

Work to follow professional standards, when available, that are most appropriate for the task at hand, departing from these only when ethically or technically justified.

Strive to fully understand the specifications for software on which they work.

Ensure that specifications for software on which they work have been well documented, satisfy the users' requirements and have the appropriate approvals.

Ensure realistic quantitative estimates of cost, scheduling, personnel, quality and outcomes on any project on which they work or propose to work and provide an uncertainty assessment of these estimates.

Ensure adequate testing, debugging, and review of software and related documents on which they work.

Ensure adequate documentation, including significant problems discovered and solutions adopted, for any project on which they work.

Work to develop software and related documents that respect the privacy of those who will be affected by that software.

Be careful to use only accurate data derived by ethical and lawful means, and use it only in ways properly authorized.

Maintain the integrity of data, being sensitive to outdated or flawed occurrences.

Treat all forms of software maintenance with the same professionalism as new development.

3. Describe a situation in which a software engineer found data that had been saved illegally, and which may provide the engineer an advantage over his teammates. What does the code of ethics say in such cases? How would you behave in such a case?
4. Describe a situation in which a software engineer found out just before a project's completion that inaccurate data was used. The chances that the client will find out this fact are small. If the parts of the software that are dependent on the inaccurate data are redeveloped, the software will be shipped with at least two months' delay. What does the code of ethics say in such cases? How would you behave in such a case?
5. In many situations, when time presses, programmers tend to skip tests. As a project leader, how would you avoid such situations?
6. eXtreme Programming (see Chapter 2, "Software Engineering Methods") addresses the activity of testing very systematically. Read about the eXtreme Programming way of testing and explain how it ensures that test will not be skipped.

TABLE 5.6 Software Engineering Code of Ethics and Professional Practice—Principle 4, Judgment

Principle 4 JUDGMENT Software engineers shall maintain integrity and independence in their professional judgment. In particular, software engineers shall, as appropriate:

Temper all technical judgments by the need to support and maintain human values.

Only endorse documents either prepared under their supervision or within their areas of competence and with which they are in agreement.

Maintain professional objectivity with respect to any software or related documents they are asked to evaluate.

Not engage in deceptive financial practices such as bribery, double billing, or other improper financial practices.

Disclose to all concerned parties those conflicts of interest that cannot reasonably be avoided or escaped.

Refuse to participate, as members or advisors, in a private, governmental or professional body concerned with software related issues, in which they, their employers or their clients have undisclosed potential conflicts of interest.

QUESTIONS—PRINCIPLE 4: JUDGMENT

1. Suggest a situation in software development in which a software developer faces a dilemma whether to consult to a company that may use her advice for competing with her own company. The software developer will earn a lot of money for this consultation. The chances that her advice will be used against her company are not clear. What does the code of ethics say in such cases? How would you behave in such a case?
2. Suggest a situation in software development in which a software developer is asked to develop a gambling game that leaves small winning chances to the gambler (relative to the accepted percentage in such cases). The software developer will get a nice salary for this work. What does the code of ethics say in such cases? How would you behave in such a case?
3. Your software company is asked to develop a system for an employment agency. The client asks that different salaries be offered to different sectors (for example, male applicants will be offered salaries different from those offered to female applicants). It is obvious that the client is attempting to check how far your company will allow discrimination. If the client observes that you accept this requirement, further features that discriminate between applicants will be requested. You and your team are asked to express your opinion. Refer to the following questions:
   - What principles of the code of ethics (if any) are violated by the customer's request?
   - Are there circumstances under which such a software product is acceptable?
   - If the customer does not agree to change their requirements, how would you react?
4. The government of a developing country wants to promote the country by offering access to the Web to all its citizens. The government faces several problems. First, since most of the population does not have the needed financial resources, at the beginning of the process only the rich people will be able to use it. Second, several sectors object to this initiative for religious reasons. Third, several private sectors admit that they will not support this initiative, as they want it to be a private project (not the government's initiative). The government knows that this initiative is the only way to join the developed part of the world.
   - What ethical issues are raised in this case? How does the code of ethics address these issues?
   - If you were a member of this government, how would you proceed?

TABLE 5.7 Software Engineering Code of Ethics and Professional Practice—Principle 5, Management

Principle 5 MANAGEMENT Software engineering managers and leaders shall subscribe to and promote an ethical approach to the management of software development and maintenance. In particular, those managing or leading software engineers shall, as appropriate:

Ensure good management for any project on which they work, including effective procedures for promotion of quality and reduction of risk.

Ensure that software engineers are informed of standards before being held to them.

Ensure that software engineers know the employer's policies and procedures for protecting passwords, files and information that is confidential to the employer or confidential to others.

Assign work only after taking into account appropriate contributions of education and experience tempered with a desire to further that education and experience.

Ensure realistic quantitative estimates of cost, scheduling, personnel, quality, and outcomes on any project on which they work or propose to work, and provide an uncertainty assessment of these estimates.

Attract potential software engineers only by full and accurate description of the conditions of employment.

Offer fair and just remuneration.

Not unjustly prevent someone from taking a position for which that person is suitably qualified.

Ensure that there is a fair agreement concerning ownership of any software, processes, research, writing, or other intellectual property to which a software engineer has contributed.

Provide for due process in hearing charges of violation of an employer's policy or of this Code.

Not ask a software engineer to do anything inconsistent with this Code.

Not punish anyone for expressing ethical concerns about a project.

QUESTIONS—PRINCIPLE 5: MANAGEMENT

1. Suggest two situations in software development that deal with password policies and procedures for protecting information, for which a manager may use the Code of Ethics of Software Engineering. Do these cases raise a conflict between the manager's and the employee's point of view? In general, can you

characterize cases in which the management and the employees may have different perspectives at a given situation? If such situations exist, how might they be solved?
2. Suggest a situation in software development in which a software developer expresses ethical concerns with respect to the leading project of the company. Describe four possible reactions on the part of the management to these concerns: two of these reactions stand in line with the Code of Ethics of Software Engineering, and two reactions conflict with the Code of Ethics of Software Engineering.
3. Since software products influence so many domains, managers of software houses face many ethical dilemmas. Here are several examples: how to maximize the company profit and at the same time develop only software tools that contribute to the society? Should only ethical considerations influence management decisions even when such considerations may hurt the company? How does management ensure that the employees do not suffer from the management effort to maximize profits? Furthermore, in considering how to behave, management has to consider the different stakeholders of their company, such as clients, shareholders, employees, partners, suppliers, environment, community, and even the state. You are asked to select at least two stakeholders and to suggest a situation in which management faces an ethical dilemma that may hurt one of the parties no matter what decision is made. How would you behave in such case? Interview several of your friends about this case. What is their reaction?

QUESTIONS—PRINCIPLE 6: PROFESSION

1. Suggest a situation in software development in which a team of software developers tries to establish the code of ethics as the norms of behavior in their company, but they face resistance on the part of the other engineers. How would you suggest they proceed?
2. Discuss connections between this section of the code of Ethics of Software Engineering and the discussion about learning organization (see Chapter 10, "Learning Processes in Software Engineering").

TABLE 5.8 Software Engineering Code of Ethics and Professional Practice—Principle 6, Profession

Principle 6 PROFESSION Software engineers shall advance the integrity and reputation of the profession consistent with the public interest. In particular, software engineers shall, as appropriate:

Help develop an organizational environment favorable to acting ethically.

Promote public knowledge of software engineering.

Extend software engineering knowledge by appropriate participation in professional organizations, meetings and publications.

Support, as members of a profession, other software engineers striving to follow this Code.

Not promote their own interest at the expense of the profession, client or employer.

Obey all laws governing their work, unless, in exceptional circumstances, such compliance is inconsistent with the public interest.

Be accurate in stating the characteristics of software on which they work, avoiding not only false claims but also claims that might reasonably be supposed to be speculative, vacuous, deceptive, misleading, or doubtful.

Take responsibility for detecting, correcting, and reporting errors in software and associated documents on which they work.

Ensure that clients, employers, and supervisors know of the software engineer's commitment to this Code of ethics, and the subsequent ramifications of such commitment.

Avoid associations with businesses and organizations which are in conflict with this code.

Recognize that violations of this Code are inconsistent with being a professional software engineer.

Express concerns to the people involved when significant violations of this Code are detected unless this is impossible, counter-productive, or dangerous.

Report significant violations of this Code to appropriate authorities when it is clear that consultation with people involved in these significant violations is impossible, counter-productive or dangerous.

TABLE 5.9 Software Engineering Code of Ethics and Professional Practice—Principle 7, Colleagues

> **Principle 7 COLLEAGUES** Software engineers shall be fair to and supportive of their colleagues. In particular, software engineers shall, as appropriate:
>
> Encourage colleagues to adhere to this Code.
>
> Assist colleagues in professional development.
>
> Credit fully the work of others and refrain from taking undue credit.
>
> Review the work of others in an objective, candid, and properly-documented way.
>
> Give a fair hearing to the opinions, concerns, or complaints of a colleague.
>
> Assist colleagues in being fully aware of current standard work practices including policies and procedures for protecting passwords, files and other confidential information, and security measures in general.
>
> Not unfairly intervene in the career of any colleague; however, concern for the employer, the client or public interest may compel software engineers, in good faith, to question the competence of a colleague.
>
> In situations outside of their own areas of competence, call upon the opinions of other professionals who have competence in that area.

QUESTIONS—PRINCIPLE 7: COLLEAGUES

1. Suggest a situation in software development in which a software developer has to decide whether to take an undue credit and get a promotion and a bonus, or to admit her actual contribution to a project and stay at the same position in the organization. What does the code of ethics say in such cases? How would you behave in such a case?
2. Suggest a situation in software development in which the head of a software team has to decide whether to accept a new opinion that seems relevant for the current project he manages. Acceptance of this opinion implies that he has managed the project incorrectly so far. What does the code of ethics say in such cases? How would you behave in such a case?
3. Assume that you and one of your colleagues are candidates for promotion to team leader, but are still members of the same team. How might this knowledge influence your behavior? How might it influence your colleague? How would you behave if one of your colleagues behaves in a way that does not stand in line with the code of ethics?

TABLE 5.10 Software Engineering Code of Ethics and Professional Practice—Principle 8, Self

Principle 8 SELF Software engineers shall participate in lifelong learning regarding the practice of their profession and shall promote an ethical approach to the practice of the profession. In particular, software engineers shall continually endeavor to:

Further their knowledge of developments in the analysis, specification, design, development, maintenance and testing of software and related documents, together with the management of the development process.

Improve their ability to create safe, reliable, and useful quality software at reasonable cost and within a reasonable time.

Improve their ability to produce accurate, informative, and well-written documentation.

Improve their understanding of the software and related documents on which they work and of the environment in which they will be used.

Improve their knowledge of relevant standards and the law governing the software and related documents on which they work.

Improve their knowledge of this Code, its interpretation, and its application to their work.

Not give unfair treatment to anyone because of any irrelevant prejudices.

Not influence others to undertake any action that involves a breach of this Code.

Recognize that personal violations of this Code are inconsistent with being a professional software engineer.

QUESTIONS—PRINCIPLE 8: SELF

1. Suggest three activities that can be done on a regular basis to fulfill the "Self" section of the Code of Ethics of Software Engineering.

Table 5.11 presents the signature of the code, including the credit to its developers.

TABLE 5.11 Software Engineering Code of Ethics and Professional Practice—Code Signature

> This Code was developed by the IEEE-CS/ACM joint task force on Software Engineering Ethics and Professional Practices (SEEPP):
>
> Executive Committee: Donald Gotterbarn (Chair), Keith Miller and Simon Rogerson;
>
> Members: Steve Barber, Peter Barnes, Ilene Burnstein, Michael Davis, Amr El-Kadi, N. Ben Fairweather, Milton Fulghum, N. Jayaram, Tom Jewett, Mark Kanko, Ernie Kallman, Duncan Langford, Joyce Currie Little, Ed Mechler, Manuel J. Norman, Douglas Phillips, Peter Ron Prinzivalli, Patrick Sullivan, John Weckert, Vivian Weil, S. Weisband and Laurie Honour Werth.
>
> ©1999 by the Institute of Electrical and Electronics Engineers, Inc. and the Association for Computing Machinery, Inc.

DISCUSSION

In this chapter, we reviewed the Code of Ethics of Software Engineering. This was done mainly by a focused analysis of case studies. As can be observed, many of the scenarios involve a combination of financial, technical, and human ethical considerations. In most of the cases, a conflict of interest should be solved.

- Is the Code of Ethics of Software Engineering different from codes of ethics of other professionals? In what sense?
- To what other (kinds of) professions might the Code of Ethics of Software Engineering be appropriate?
- Is it possible that two principles of Code of Ethics of Software Engineering contradict each other? If yes, give an example of such a case. How can the conflict be resolved? If not, explain why.

SUMMARY QUESTIONS

1. What topics are addressed in the code of ethics?
2. Describe in two to three sentences the essence of each section of the Code of Ethics of Software Engineering.
3. In the "Relevance for Software Engineering" section, we said that although the code of ethics can be addressed on different levels (individual, team, organization, country), in this chapter we limit the discussion to the individual level. At this stage, you are asked to suggest situations from the daily life

of software developers in which ethical considerations address the team, the organization, and the country levels.

FOR FURTHER REVIEW

1. Visit a software house and ask software developers about dilemmas they faced that required them to address ethical considerations. Find out if they are familiar with the Code of Ethics of Software Engineering. Ask them to analyze the situations they describe by using the Code of Ethics of Software Engineering.
2. The Web is full of stories and case studies that raise ethical dilemmas. See the end of this chapter for a partial list. Select two to three stories and for each case check in what ways the code of ethics is helpful. Specify what principles of the Code of Ethics of Software Engineering you considered.
3. Compose a story that raises ethical considerations. Interview software engineers about this case. Ask them to express their opinion and behavior in such a case. Analyze these reactions. Are all of the reactions similar? In what ways do they differ from each other and from your opinion? What do these reactions imply with respect to software development? What lessons will you take from this experience in your future development of software?
4. Intellectual property in the age of computing is a central topic when analyzed from the ethical perspective. Surf the Web, find basic ethical rules that address this topic, and connect them to the daily life of software engineers.
5. Netiquette is the ethics of using e-mail and other types of electronic communication. Search the Web for the basic rules of Netiquette. Select several of its guidelines, and explain their source and importance.

REFERENCES AND ADDITIONAL RESOURCES

The Code of Ethics of Software Engineering: *www.computer.org/tab/seprof/code.htm*.

[Kasher02] Kasher, Asa, "Professional Autonomy and Its Limits," *The Research in Ethics and Engineering* conference (April 25–27, 2002), Delft University of Technology, *www.ethiek.tudelft.nl/conference2002/abstracts.htm#kasher*.

Bowen, Jonathan P., "The Ethics of Safety-Critical Systems," *Communications of the ACM* 43(4), (April 2000): pp. 91–97.

Collins, Robert, W., et al., "How Good is Good Enough? An ethical analysis of software construction and use." *Communications of the ACM* 37(1), (January 1994): pp. 81–91.

Johnson, Deborah, *Computer Ethics,* Third Edition, Prentice Hall, 2001.

Kreie, Jennifer, and Cronan, Timothy Paul, "Making Ethical Decisions," *Communication of the ACAM*, 43(12) (December 2000): pp. 66–71.

Towell, Elizabeth, "Teaching Ethics in Software Engineering Curriculum," *Proceedings of the 16th conference on Software Engineering Education & Training*, Madrid, Spain, (2002): pp. 150–157.

Online Resources

The Research Center on Computing & Society at Southern Connecticut State University, Computer Ethics in the Computer Science Curriculum, *www.southernct.edu/organizations/rccs/resources/teaching/teaching_mono/bynum/bynum_capstone.html.*

The National Institute for Engineering Ethics: *www.niee.org/cases.* The Web site also presents an analysis of the case study.

CPSC 451: Practical Software Engineering course of the University of Calgary. *http://sern.ucalgary.ca/courses/cpsc/451/W99/Ethics.html#case1.* The Web site also presents an analysis of the case studies.

The Software Engineering Ethics Research Institute: *http://seeri.etsu.edu/.*

Professor Richard Upchurch, University of Massachusetts Dartmouth, Computer and Information Science, Social and Ethical Aspects of Computing course Web site: *www2.umassd.edu/CISW3/coursepages/pages/cis381/outline.html.*

Nancy Leveson, Forum on Risks to the Public in Computers and Related Systems, *ACM Committee on Computers and Public Policy,* Peter G. Neumann, moderator, Volume 6: Issue 18, January 29, 1988. The forum URL is *http://catless.ncl.ac.uk/Risks/6.18.html#subj1.*

ENDNOTES

[1] The Code can be found at *www.acm.org/constitution/code.html.* It is also presented in this chapter.

[2] The code can be published as is, as it is stated clearly that "[t]his Code may be published without permission as long as it is not changed in any way and it carries the copyright notice." These two requests are fulfilled in this book.

6 International Perspective on Software Engineering

In This Chapter

- Introduction
- Objectives
- Study Questions
- Relevance for Software Engineering
- International Perspectives on Software Engineering
- The High-Tech Industry in Different Countries
- Additional International Topics Related to Software Engineering
- Women and Minorities in Computer Science and Software Engineering
- Summary Questions

INTRODUCTION

In this chapter, the field of software engineering is examined from an international perspective. In a way, the scope of this chapter is similar to the scope of Chapter 5, "Code of Ethics of Software Engineering." In both chapters, the focus is on the community of software engineers beyond the team and the organization framework. Specifically, in this chapter, we explore the influence of certain events on the global high-tech economy and the nature of software engineering in several places on the globe. In addition, we discuss related topics, such as gender and minorities in the high-tech industry.

OBJECTIVES

- Readers will become aware of the potential influence of local events on the global high-tech industry.
- Readers will become familiar with the characteristics of software engineering processes in different places around the world.
- Readers will increase their awareness of the influence of different cultures on software engineering processes.
- Readers will become familiar with topics related to gender and minorities in the high-tech culture.

STUDY QUESTIONS

1. What characteristics make software engineering a suitable topic to be discussed from an international perspective?
2. Select three countries and search the Web for information about their high-tech economy. In what ways is the high-tech culture in these countries similar? In what ways is it different?
3. If you had the option to establish a software house anywhere on the globe, what considerations would you address during the decision-making process? How would each of these considerations influence your decision?
4. Assume that your company has two sites in two different countries. How will you share development between the two sites? What communication channels between the two sites will you establish?
5. After the September 11 terror attack on the United States, the Pew Internet & American Life Project published the following report: "The commons of the tragedy: How the Internet was used by millions after the terror attacks to grieve, console, share news, and debate the country's response."
   Read the report at *www.pewinternet.org/reports/pdfs/PIP_Tragedy_Report.pdf*. How would you explain the facts it presents?
6. Identify three additional events that had a global influence on the high-tech industry. Describe the influence of each event.

RELEVANCE FOR SOFTWARE ENGINEERING

Since many software development companies are international firms, it is important that software engineers be aware of different cultures in general and of different cultures of software development in particular. A general discussion about cultural differences is beyond the scope of this book. Thus, this chapter focuses on two countries—India and Israel—that are unique in some sense with respect to the software market. Specifically, we describe cultural factors that may explain the success of the high-tech industry in these countries. Our aim is to highlight the fact that software development may be influenced by local and cultural characteristics. We believe that the more software engineers are aware of such topics and consider them when they cooperate with people from other countries and/or cultures, the more fruitful their software development processes become.

A topic closely related to multiple cultures in the information technology professions is gender and minorities. As it turns out, in many countries women are underrepresented in the high-tech field. Since a lot of research has been conducted so far with respect to this topic, in this chapter we present the main findings of the research and present reasons that may explain this situation. The relevance of this topic to software engineering is as follows: because the software engineering area is still evolving, the people who participate in its development may have a significant influence on its future. If our intention is to expand the population who uses software products, it is reasonable to assume that the more diverse the developer population becomes, the bigger the population that software products fit will become.

In this chapter, we do not merely focus on software engineering, but examine the topic from the wider perspective of the Information Technology (IT) market, and sometimes even from the perspective of the entire high-tech industry. It is clear, however, that the impact of the topics discussed in this chapter with respect to the IT and high-tech fields have direct implications on software engineering processes.

INTERNATIONAL PERSPECTIVES ON SOFTWARE ENGINEERING

The timing of writing this chapter is significant. It is written after the Internet bubble burst and after the world experienced two events that dramatically influenced the high-tech economy—the September 11, 2001, terrorist attacks in the United States, and the SARS (Severe Acute Respiratory Syndrome) epidemic in Asia. In this spirit, this chapter illustrates how the global high-tech industry is in fact one worldwide market. We start by showing how the September 11 terror attack and the SARS outbreak influenced the high-tech economy far beyond the regions where they happened. In both cases, the international markets in general and the high-tech markets in particular suffered.

On September 11, 2001, the United States experienced a terrorist attack. That day is described in terms such as "tragedy" and "disaster." Naturally, our discussion is limited to the influence of that day on the high-tech economy. Any such event leads to a chain of behaviors, each of which increases the negative influence of the event. In the case of September 11, people tended to fly less than they did before, purchases were reduced, and the entire economy slowed.

The September 11 terrorist attack raised many questions about computer network security and other channels of information sharing. Some companies took advantage of this failure in information gathering and the increased need for protecting information. In addition, the need has emerged for tools to analyze the gathered information. This fact illustrates that even disasters can promote the development of technology and open business opportunity.

SARS is a disease that makes it difficult to breathe. In contrast to a bad cold, it can kill people if it is not treated quickly. It is suspected to have originated in Guangdong, China, in November 2002. However, China was not the only country that suffered from SARS; other places, such as Singapore and Canada, also had many cases of SARS. As it turns out, SARS dramatically influenced the high-tech economy (which was going through a deep recession in those days anyway), mainly because people were afraid to travel to those regions. Consequently, during the outbreak, many companies canceled any non-mission-critical business travel to Asia.

The influence of SARS was dramatic, since Asia is one of the fastest-growing regions in the world today. Furthermore, many companies outsource their manufacturing to China because of the perceived lower cost, high quality, and dependability. Following the SARS epidemic, companies started reexamining their outsourcing plans in general and whether they could put all their high-tech risks in the China basket in particular.

Questions

1. Both September 11 and SARS influenced the high-tech economy negatively. Suggest events that may have a positive influence on the high-tech economy.
2. Among other factors that influenced the high-tech market globally, we find the new browsers that emerged in the early 1990s and enabled each person to "surf the Web" easily without having any technical background. In addition, new e-commerce Web sites enabled trading between countries without borders and tax limitations. Suggest additional ways by which the Web may influence the world globally. Refer to different areas of influence (medicine, education, etc.).
3. Look at Web sites that present Internet statistics and find data about the number of people who have Internet access in different places on the globe. What do these numbers reflect with respect to the global economy?
4. Linux and other Open Source software are concepts that have been mutually influenced by the global high-tech networking. How have these concepts been influenced by the fact that the globe became one online village? Look for information about the evolution of these concepts.

THE HIGH-TECH INDUSTRY IN DIFFERENT COUNTRIES

This section examines the high-tech industry from the country-level perspective and looks at some interconnections between the high-tech industry in different places.

In the last decade of the twentieth century the high-tech sector flourished. Advertisements for IT workers were published all over the world, and headhunters could not fill the huge demand for these workers. Thus, not only were products spread worldwide, development itself also became an international process. This trend dramatically increased the interconnections among countries with respect to software development.

The huge demand for IT workers characterized both developed and developing countries. The former needed IT workers for research and development and for maintaining the growing high-tech sector; the latter needed IT workers for the creation of their IT infrastructure. In general, the number of IT workers did not answer the demand, and even new education programs could not cope with this challenge. As with other sectors, where the demand increases the supply, this competition led to an increase in salaries, work conditions, and Visa permissions (in the case of the US H1B visas). Immigration to countries such as the United States, the United Kingdom, Canada, Japan, Ireland, and Australia was encouraged.

TASK

At the beginning of the 1990s, Edward Yourdon wrote *Decline and Fall of the American Programmer* [Yourdon92]. The book assesses the way the American software industry behaves in the global marketplace. Based on this book and other resources, explore similarities among the phenomenon described in the book with respect to the software industry and the processes through which the U.S. automobile industry went several decades ago.

Two main factors determined why IT experts preferred moving to the United States in the 1990s. The first is the huge salary gap between the United States and other (western as well as not western) countries. These salaries enabled many IT workers to move to the United States and support families who stayed in the homeland. The second factor was the attractive cutting-edge technology that the IT experts had an opportunity to work with in the United States.

During the high-tech flowering of the 1990s the press and other media channels kept reporting about amazing high-tech stories in the technology market. Many young people without any business experience entered the area hoping to earn money quickly. Many investors were willing to invest money in these adventures. That new atmosphere changed traditional work relationships. Innovators and skilled programmers were sought out, and spoiled with good benefits. These benefits caused people to move from one software house to another according to who paid more. When this happens to young people, it is not a healthy phenomenon.

The year 2000 was a turning point in the high-tech market. When it became clear that some companies did not have any revenues, the investors stopped the money stream and those companies failed. The exaggerated expectations of IT products were misleading and a surplus emerged. This led naturally to a crisis and the stock exchange rate decreased dramatically.

TASKS

These tasks examine the NASDAQ (National Association of Securities Dealers Automated Quotation) during the 10-year period from 1993 through 2003.

1. Select five years during this period. For each year, find what countries had companies listed on the NASDAQ. What does this list of countries say about the NASDAQ and about the international market in these years?
2. Conduct a close examination of the years 1999, 2000, 2001, and 2002. Select four months in each year and compare the NASDAQ rate in these months. What trends can you observe? How can you explain them?

The year 2000 economic crisis in the United States spread to the entire world. The demand for IT workers decreased dramatically, yet many IT companies fired their workers and others went out of business. A major part of this economic downturn is known as the "dot-com crisis." This crisis refers to the collapse of the many startup companies that dealt with Internet products, their security, e-commerce applications, and so forth. Looking back at these years, that collapse is not surprising at all: the overbudget expenses for the IT workers, the absence of focus on management and vision, neglect of customer satisfaction, and unprofessional development together led very naturally to the crash of the Internet bubble in the spring of 2000.

TASK

Tell a story about a typical dot-com company that eventually collapsed. Focus on how its management led to its collapse together with other dot-com companies. What would you change in its management to avoid its collapse? Think about factors such as lack of a plan and vision, inappropriately high salaries, and high expenses without control. You can look at the Web to find hints about such stories.

The year 2001 was the worst for the computer industry all over the world. Some people talked about 2002 as the year of the recovery. However, until this writing (2004), these hopes for recovery are still pale. The high-tech market is far behind the market of the 1990s. The extravagant period is over; the world seems to have awakened from the high-tech dream.

In each country, the software industry is tightly connected to other domains. The government decides on foreign policies (such as work visa and import and export policies), legal issues (such as taxes), and budget allocations to different sectors (such as R&D and education). The industrial sector determines what types of products are developed, whether development is emphasized over research, and so forth. Academia sets the education level and the teaching methods and establishes connections with industry. Finally, factors related to the population of each country matter as well: How many people are educated? What are the average living standards? What are the culture and traditions? How might culture and tradition influence the software industry? What is the age distribution of the population? Is the population aware of the importance of education? These aspects are addressed when we discuss the IT field in two specific countries—India, a large target of American outsourcing, and Israel, known very well by one of the authors.

TASK

Suppose that you are the person in charge of the budget allocation in your country and you should decide on the priorities among education, local industry, and foreign policy (e.g., attraction of workers from abroad). What percentage of the total budget would you allocate to each sector? Since there is no magic formula to apply in such situations, try to determine the best budget allocation for your country, taking into consideration its specific and unique characteristics and needs.

India

TASKS

1. Search the Web for three Indian high-tech companies. Describe their products and the way they present themselves. Is there a difference between the way these companies are presented and the image of American companies? If yes, in what sense? If not, what is common to the images that are presented in the two cultures?
2. Find what high-tech giants usually thought of as American established a site in India. Explain the motives for such a business action.

India got its independence from Great Britain on August 15, 1947, so it is relatively new as an independent country. Its population is currently thought to be more than 1 billion people (the second most populous in the world after China), and it has the second highest number of English speakers after the United States. This last fact ought to give India a big advantage in the high-tech market. As a result, it became a huge source for high-tech workers. A notable fact is that despite the worldwide recession, India's software and services exports rose by 28 percent to 6.9 billion U.S. dollars in the nine months ended December 2002 (*www.indiatravelogue.com/pass/pass4.html*).

India has 4,500,000 IT workers, and the high-tech industry has become a central factor in the development of the Indian economy. As mentioned previously, India is a huge resource for high-tech workers, and many international companies hire Indian software developers. Sometimes companies open a new site in India, and sometimes the Indian workers move to other countries. The Indian government understands the potential of this sector and encourages both the establishment of new IT schools (to educate the next generations) and the growth of the existing IT educational institutions. Although the many IT schools seem far behind with technology, they are far ahead with students' motivation to learn the new high-tech trends and programming languages.

As mentioned before, the government also encourages foreign companies to invest in India and to open new development sites. This is done, for example, by reducing bureaucracy. In addition, the government supports local development by allocating relatively huge budgets for this purpose. This support is also expressed by canceling the taxes on the export of technological products and technological services. All of these actions together increased dramatically the business interest that Western companies found in India. Consequently, during the last decade India became an important player in the global IT economy.

TASKS

1. The structure of Indian society is hierarchical and consists of castes. Some think that this tradition led to the huge social gaps that exist in India. In your opinion, how will this social structure survive in the high-tech era that India enters into?
2. Similar to India, other Eastern countries have discovered their IT potential and can offer a cheap and skilled workforce. One of these countries is China. In your opinion, what advantages and disadvantages does China have over India, and vice versa? What processes that deal with the high-tech economy might the two countries experience in the next decade?

Israel

TASK

Search the Web for three Israeli high-tech companies. Describe their products and the way in which they present themselves. Is there a difference between their image and the image of American companies? If yes, in what sense? If not, what is common to the profiles.

Israel got its independence from Great Britain on May 15, 1948, making it slightly younger than India. There the similarities stop. It is a very small country and its population is about 6.5 million people. At its economic peak, Israel was considered a high-tech center and even imported software engineers from India. These differences in size and similarities in the high-tech orientation of India and Israel were two of the reasons we selected these two countries for close examination in this chapter.

As mentioned earlier, in its high-tech peak Israel was one of the leading centers for technology startups and innovations. Despite its small population, Israel had about 3,000 startups (*www.ite.poly.edu/htmls/role_israel0110.htm*). In that era, Israel had the third (after the United States and Canada) highest number of companies listed on the NASDAQ. This situation led naturally to the establishment of many educational programs and, as mentioned before, to the importation of programmers from other countries to Israel.

This blossoming is explained by two main factors. The first is the national security and military needs that led to the development of cutting-edge technologies, in particular, designated army units that specialize in technology innovations. As it turns out, many of the Israeli's high-tech entrepreneurs started their careers in the Israel Defense Forces.

The second factor that explains the Israeli high-tech success in the 1990s is the huge immigration of Russian engineers from the former Soviet Union during the decade of 1990 to 2000. Again, we see how one event, in one place on the globe (in this case, the open gates of the former Soviet Union) influences the high-tech economy elsewhere. This addition of engineers to the Israeli population led Israel to have the highest number of engineers per capita in the world (*www.smartcodecorp.com/about_us/israel_profile.asp*).

The peak of this period was the year 2000. Foreigners invested billions of U.S. dollars in Israeli startups and venture capital funds. In that year, the high-tech exports reached record highs and comprised 57 percent of the entire export of Israel. As in the case of India, the Israeli government had a significant role in this progress. It invested many U.S. dollars in research and development, in the establishment of scientific incubators, and in other enterprises.

Many companies founded in Israel during that period moved their headquarters to the United States, where much of their customer base is located, but often left their research and development operations in Israel. Some of these companies were purchased by American companies. For example, Mirabilis (the inventor of ICQ—I Seek You) was purchased in June 1998 by AOL for $287 million (*www.ccta.ca/english/publications/submissions/1999-98/1998-10-01.htm*). In May 2000, Chromatis (which developed optical networking) was purchased by Lucent Technologies for about $4.5 billion US (and was closed by Lucent in August 2001) (*www.lucent.com/press/0500/000531.coa.html*). The strength of the Israeli software firms was and still is innovative and daring R&D. Not surprisingly, IBM, Intel, and Microsoft have research and development centers in Israel.

Because of the globalization process, Israel, together with the entire world, suffers now from a deep recession. To illustrate the huge change in the Israeli high-tech economy before and after year 2000, here is some data taken from the Israeli National Bureau of Statistics: high-tech production in 2001 decreased by 23 percent, after a sharp increase of 26 percent of production in 2000! The demand for high-tech workers in the fourth quarter of 2001 decreased by 83 percent relative to the same quarter a year earlier. Not surprisingly, salaries decreased.

Currently, Israel joins in the worldwide desire for recovery. However, one bright light still exists in the Israeli high-tech economy—the biotechnology sector. This sector is comprised of 160 biotechnology firms with about 4,000 workers, and is performing relatively well. (*www.mfa.gov.il/mfa/go.asp?MFAH0mun0*).

TASK

Select two countries and describe their IT industry in the same way we described India and Israel.

ADDITIONAL INTERNATIONAL TOPICS RELATED TO SOFTWARE ENGINEERING

So far, we looked at software engineering industry in the United States, India, and Israel. Naturally, these are only three countries and there is more to say about what goes on globally with respect to software engineering. Among other related topics, there are attempts to help the developing countries become more technology oriented. For example, there is a UN initiative to help developing countries boost IT infrastructure, especially in countries that are lagging behind the rest of the world in these areas. The idea is to grant these countries seed money programs to help them get started. It is accepted that each country will continue after that in the development processes with the support of other governments and companies.

There is consensus that the problem of spurring the developing countries can be solved mainly by their educational systems. To help governments with this challenge, the private sector should take a more active role in extending the opportunities afforded through information technology to the poor and disadvantaged parts of the world. Indeed, many companies initiated special programs to promote their relationships with the community. Intel's Innovation in Education project is just one example[7] (*www.intel.com/education/projects/global_tour/*).

Part of the gaps can also be reduced by equal opportunity. As a starting point, some governments have to address historical problems of gender discrimination and other equality issues in order to help all their citizens take advantage of the knowledge economy and to help the underprivileged become knowledgeable workers.

We sum up this section of this chapter thus:

1. The last decade can be conceived of as a decade of the natural selection of high-tech companies that match economic conditions (similar to the biological evolutionary process). Accordingly, when the worldwide high-tech industry recovers, the main lesson investors will remember is to invest money proportionally to what companies offer and not based on their optimism.
2. The development of the high-tech markets enables developing countries to enter the high-tech industry. With their relatively cheap workforce, some of these countries may gradually acquire position and influence the high-tech sectors.
3. The IT niche directly influences areas such as education, economy, diplomacy, unemployment, and so forth. Specifically, it becomes clear that an educated population is a central condition for international economical success.
4. The high-tech economy may serve as a means for closing cultural and social gaps.

WOMEN AND MINORITIES IN COMPUTER SCIENCE AND SOFTWARE ENGINEERING

Data indicates that both women and minorities are underrepresented in the community of IT developers. However, in this section we focus on the issue only as it is expressed with respect to women. Let us explain this choice. First, women comprise about 50 percent of the world population and account for half of the current workforce, yet are only 20 percent of the technology professionals (*www.acm.org/technews/articles/2003-5/0321f.html#item3*). Second, the absence of women in the high-tech industry is almost a worldwide phenomenon, at least in the Western countries. Third, as each country has its own minorities, and it is reasonable to assume that local factors influence the level of their involvement in the IT economy, it would be impossible to deal with all these local differences. Even though the focus in this section is on the under-representation of women in the IT workforce, we believe that based on this discussion, readers will be able to make relevant conclusions with respect to minorities.

In the discussion that follows, we do not explicitly distinguish between computer science, software engineering, and the other IT professions. In some contexts, the data we present refers to the study of computer science; in other cases, the data describes a phenomenon related to women in the IT professions in general. However, as it turns out, one message is reflected by all these numbers. The percentage of women who enter these areas is relatively low, and some action should be taken to increase the presence of women in these professions. Indeed, the topic of women in the IT field currently gets a lot of attention. Still, real impact and change have not happened yet.

In research published in 1997 by Tracy Camp [Camp97] and later in research published by Vanessa Davies and Tracy Camp in 2000 [Davies and Camp00], a phenomenon called "the shrinking pipeline" was identified. This phenomenon addresses the fact that the number of women who studied toward a bachelor degree in computer science decreased from 37 percent in 1983 to about 28 percent a decade later and to 15 to 20 percent in the second half of the last decade. In addition, the percentage of women who hold higher positions in the universities decreases the higher the position is. A similar phenomenon has been identified in other countries as well.

These findings led many organizations (especially academic ones) to take action to increase the percentage of women who study computer science. For example, starting in 1995, in the Carnegie Mellon University School of Computer Science, an interdisciplinary program of research and action was initiated. The research aspect of this project aimed to understand students' attitudes toward computer science; the action component aimed to devise and effect changes in curriculum, pedagogy, and culture that will encourage the broadest possible participation of women in the school. As it turns out, all these efforts have raised the entering enrollment of women in the undergraduate computer science program at Carnegie Mellon University from 8 percent in 1995 to 42 percent in 2000. The full story of this project is described in "Unlocking the Clubhouse: Women In Computing" [Margolis and Fisher02].

A relevant question is "why bother?" In other words, why is the underrepresentation of women in the IT professions considered a problem? We present three arguments that explain the importance of increasing the number of women in the IT sector in general, and in the computer science and software engineering professions in particular. First, there is the general issue of female equity in opportunities. Because these professions promise, in most cases, high salaries, women should share in them. Second, as computer science and software engineering are relatively young disciplines, if women become part of the professional community, they may influence the way these fields are shaped. Finally, sometimes computational systems are designed in a way that do not fit women's needs. If women take an active role in the design of these systems, it is more likely these systems will answer women's needs as well. Examples of such systems are presented in [Margolis and Fisher02].

TASKS

1. How would you explain the fact that the percentage of women who study computer science and software engineering is relatively low? What factors might influence their professional choices?
2. How would you approach the problems?

There are many reasons that lead women not to study computer science and software engineering. First, the image of these professions is that only "nerds" work in these areas, and that a typical workday is made of long hours of coding, coding, and coding, without any human interaction [Jepson and Perl02]. As a reader of this book, you must realize that this image is wrong and that human aspects are a major component of software development. However, the incorrect conception of these professions that women hold, and their tendency to prefer jobs that are based on human interaction, causes many of them not to consider studying IT-related professions at all. This contradiction can be solved in different ways. For example, at Carnegie Mellon University, as part of the project mentioned earlier, several actions in this direction took place. Among other activities, an undergraduate "immigration course" was instituted. The aim of the course was to present students with a broader view of computer science than the traditional view they usually get in the other first-year courses they learn.

Another factor that influences women not to study high-tech professions is their early education and exposure to technology [Fisher and Margolis02]. To get an idea about the dominance of this fact, it is sufficient just to compare the variety of computer games that are offered to girls and to boys. Go to any computer game shop and see the huge assortment of games aimed at boys, relative to the almost absence of games offered to girls [Gürer and Camp02].

TASKS

1. Outline the main features of a computer game that in your opinion fits and may appeal to young girls.
2. Conduct the same task for a game that fits and may appeal to young boys.
3. Are there differences between the two games? If yes, what are the differences? If not, what do these types of games have in common?

A third factor that influences women not to study the high-tech professions is the lack of role models—women with whom young women can identify [Pearl, et al.02]. Role models may increase young women's attention to the fact that women can succeed in these areas. To let young women see that there are women who work and succeed in the high-tech world, several programs that pair female high-school students with mentors in the industry have been established. These programs aim to let young women experience for several days what a career in the tech fields means. The hope of these programs is that the young participants will be inspired and consider college and university technology and science programs.

TASKS

1. Anita Borg was a computer scientist who passed away on April 7, 2003 at the age of 54. Among her activities in response to the low number of women in the computer industry, Borg founded Systers, an e-mail list designed to provide women with an online community where they could network and mentor each other. Search for information about Anita Borg, her professional achievements, and her contribution to the education of women in the high-tech jobs. Summarize your findings. You can start by looking at *www.iwt.org/news/anitaborg/inmemory.htm*.
2. Make a similar search for a woman who works in the IT area in your country. How did she make it?

The following paragraphs illustrate that although in general women are under-represented in the IT fields, there are programs and educational initiatives where women take an active part.

We start with a local initiative. The second annual Lego League Rhode Island State Robotics Tournament was notable for the number of pre-teen and teenage female participants. This tournament was supported by a state grant and was intended to set up a girls-only robotics program. As it turned out, the number of girls who signed up for the tournament rose from 20 in 2001 to more than 50 in 2002. In one part of the competition student teams built a robot out of Lego blocks and pre-programmed it to perform a series of moves designed to fulfill a particular mission.

From a more global perspective, research work has shown that girls and young women learn scientific topics in a better way when they study in all-female classes [Galpin02]. The idea is to let female students be dominant in these classes and not be afraid to ask questions in front of their male classmates. Positive findings about this structure of science and technology classes have been found all over the world and particularly in some of the countries that do not suffer from the under-representation of women in the high-tech sector.

SUMMARY QUESTIONS

1. Discuss connections between the topics discussed in this chapter and the Code of Ethics of Software Engineering discussed in Chapter 5.
2. Summarize the main messages of this chapter. Reflect:
   - What is common to all these messages?
   - In what way do these messages differ?
   - Do these messages change your perspective of the field of software engineering? If yes, in what sense?
3. eXtreme Programming (XP) is a software development method presented in Chapter 2. Review it and analyze its fitness to women's way of learning, management style, and communication habits.

FOR FURTHER REVIEW

1. Suggest activities for the encouragement of women and other under-represented groups to join the high-tech jobs. If possible, conduct a short experiment. You may use one of the research tools presented in Chapter 4, "Software as a Product." Present your conclusions.
2. If you have to outline a worldwide agenda for the encouragement of women (or any other under-represented group) to enter the high-tech sector, what would be your suggestions? What would be the basic guidelines of this international agenda? What type of groups or organizations would you recruit for this task?

REFERENCES AND ADDITIONAL RESOURCES

[Fisher and Margolis02] Fisher, Allan, and Margolis, Jane, "Unlocking the Clubhouse: the Carnegie-Mellon Experience," *inroads—SIGCSE Bulletin* 34(2): Special Issue on Women and Computing (June 2002): pp. 79–83.

[Camp97] Camp, Tracy, "The Incredible Shrinking Pipeline," *Communications of the ACM* 40(10), (October 1997): pp. 103–110.

[Davies and Camp00] Davies, Vanessa, and Camp, Tracy, "Where have women gone and will they be returning," *The Computer Professionals for Social Responsibility Newsletter* 18(1), (Winter 2000): *www.cpsr.org/publications/newsletters/issues/2000/Winter2000/davies-camp.html*.

[Galpin02] Galpin, Vashti, "Women in Computing around the World," *inroads—SIGCSE Bulletin* 34(2): Special Issue on Women and Computing (June 2002): pp. 94–100.

[Gürer and Camp02] Gürer, Denis, and Camp, Tracy, "An ACM-W Literature Review on Women in Computing," *inroads—SIGCSE Bulletin* 34(2): Special Issue on Women and Computing (June 2002): pp. 121–127.

[Jepson and Perl02] Jepson, Andrea, and Perl, Teri, "Priming the Pipeline," *inroads—SIGCSE Bulletin* 34(2): Special Issue on Women and Computing (June 2002): pp. 36–39.

[Margolis and Fisher02] Margolis, Jane, and Fisher, Allan, *Unlocking the Clubhouse: Women in Computing*, MIT Press, 2002.

[Pearl, et al.02] Pearl, Amy; Pollack, Martha E.; Reskin, Eve; Thomas, Becky; Wolf, Elizabeth; and Wu, Alice, "Becoming a Computer Scientist," *inroads—SIGCSE Bulletin* 34(2): Special Issue on Women and Computing (June 2002): pp. 135–143.

[Yourdon92] Yourdon, Edward, *Decline and Fall of the American Programmer*, Prentice Hall, 1992.

Communications of the ACM, July 2001, Volume 44, number 7: Special Issue on the Global IT Workforce.

inroads—SIGCSE Bulletin, June 2002, Volume 34, Number 2: Special Issue on Women and Computing.

MIT Department of Electrical Engineering & Computer Science. (1995). Women Undergraduate Enrollment in Electrical Engineering and Computer Science at MIT. Introduction and Executive Summary: *www-eecs.mit.edu/AY94-95/announcements/13.html*. Full Report: *http://www-swiss.ai.mit.edu/~hal/women-enrollment-comm/final-report.html*.

Associations that Deal with the Promotion of Women in Computer Science

ACM's Committee on Women and Computing: *http://www.acm.org/women/* (additional organizations can be found under Related Sites).

The Association for Women in Computing (AWC): *http://www.awc-hq.org/*

Women@SCS at Carnegie-Mellon: *http://women.cs.cmu.edu/*

WICS—Stanford Women in Computer Science: *http://www.stanford.edu/group/wics/*

WICSE—Women in Computer Science and Electrical Engineering, University of California, Berkeley: *http://www-inst.eecs.berkeley.edu/~wicse/*

7 Different Perspectives on Software Engineering

In This Chapter

- Introduction
- Objectives
- Study Questions
- Relevance for Software Engineering
- Software Engineering: A Multifaceted Field
- Summary Questions

INTRODUCTION

This chapter explores different points of view of software engineering. It aims to highlight the fact that software engineering can be looked at and examined from different perspectives, each emphasizing a different aspect of the field. The importance of this issue stems from the fact that the discipline of software engineering is still a young and evolving field. Consequently, these different perspectives may have a significant influence on the path it takes.

Naturally, it will be impossible to review in one chapter all these perspectives of software engineering. From among the many possible topics to discuss, the following two are selected: the product versus process perspectives of software engineering [Floyd87], and the agile versus the heavyweight software development methods. We find these topics illustrative as they both emphasize the human aspects of software engineering. In addition, to illustrate the dynamic nature of the discipline and the way it evolves, we present several definitions for software engineering, taken from different stages of its development, and examine what conception of the field each reflects.

This chapter can be viewed as an introduction to an analytic approach that encourages an examination of topics related to software engineering from different angles. Chapter 11, "Abstraction and Other Heuristics of Software Development," illustrates this approach by analyzing several of the topics discussed in the book through the lens of abstraction.

OBJECTIVES

- Readers will increase their awareness that the discipline of software engineering can be viewed from different perspectives, each emphasizing different aspects of the discipline.
- Readers will be able to identify what elements from each perspective fit their perception of software engineering.

STUDY QUESTIONS

1. Chapter 2, "Software Engineering Methods," focuses on software engineering methods.
    a. Suggest different possible approaches toward this topic.
    b. Among other issues discussed in Chapter 2, the agile approach toward software development was presented. Explain this approach and discuss in what sense it differs from other approaches toward software development.
2. The topic of customer was discussed in Chapter 4, "Software as a Product." Suggest different possible approaches toward the customer's role in the software development processes.
3. Select two topics that were discussed in previous chapters of the book. Check whether they can be analyzed from different angles. To what aspects of each topic do these different perspectives refer?
4. Suggest a new topic related to software engineering that has not been discussed yet in the book. Interview several software engineers about it (see Chapter 4 for guidelines how to construct and conduct interviews). Analyze similarities and differences among the practitioners' opinions.

RELEVANCE FOR SOFTWARE ENGINEERING

Because software engineering is an evolving field, its practioners still look for one framework that will be accepted by the entire community of software engineers. Such a situation characterizes every young discipline in its early steps. Sometimes, a similar process also happens to mature disciplines. For example, there are cases in which established fields change their core research topics. In some cases, because of the conceptual change, new research methods replace methods irrelevant for the new research topics.

As a software development practitioner, it is important to be aware of different approaches that your teammates may value. It is sufficient to consider, for example, how software development processes may be immediately affected when different team members express different approaches toward topics such as the customer's role in the development process. Naturally, such different perspectives may lead to conflicts between team members from the early stages of the development process, and consequently the process may suffer.

This chapter aims to increase software engineers' awareness of the existence of different approaches to a variety of topics related to software engineering. It is further suggested that this awareness may sometimes be helpful in resolving conflicts among team members. In practice, it may guide to identify the source of the conflict and the team members' different perspectives of the discussed topic. Based on this understanding, the team can examine ways to bridge the different approaches.

SOFTWARE ENGINEERING: A MULTIFACETED FIELD

One way to convey the multifaceted nature of software engineering is to review how it is conceived of at different times since the term was coined in 1968 (see Chapter 8, "The History of Software Engineering"). Here we conduct this review by presenting several definitions of software engineering taken from three different stages of its history. The idea is to describe the shift in the focus of the discipline. More specifically, while the early definitions of software engineering from the 1970s focused on its technical aspect, later writings refer to the nature of the field as an engineering profession.

In this spirit, von Mayrhauser presents definitions for software engineering from the 1970s [vonMayrhauser90]. Among those, Zelkowitz's definition states that software engineering studies *principles* and *methodologies* for developing and maintaining software systems [Zelkowitz78] [vonMayrhauser90, p. 4]. The second definition that von Mayrhauser presents declares that software engineering is the *practical* application of scientific knowledge to the construction of computer programs and their documentation [Boehm76] [vonMayrhauser90, p. 5].

Zelkowitz's definition emphasizes the methodological aspects of the field; Boehm's definition adds and refers to the documentation of software. These definitions use terms such as *principles, methodologies, scientific knowledge, construction,* and *maintaining*. Human aspects of software engineering do not appear to be addressed explicitly in these definitions.

One of the first signs of the awareness of human aspects of software engineering appeared in Brooks' book *The Mythical Man-Month*, first published in 1975 and revised in 1995 [Brooks75]. In this book, the social complexity involved in software development and the uniqueness of this process is acknowledged. One phenomenon described in the book is that most software projects are never released on time as planned. In the preface to the *20th Anniversary Edition*, Brooks expressed surprise and delight that *The Mythical Man-Month* continued to be popular two decades after first being published. This fact emphasizes how difficult it is to apply lessons that had been learned with respect to software development. It is suggested that this difficulty may be partially explained by the multifaceted nature of the discipline and the uniqueness of the human aspect of software development processes.

As is illustrated in what follows, in present writings, the emphasis on the "engineering" aspect of software engineering is decreased. This shift in emphasis may be partially explained by the failure to formulate one engineering-oriented process that answers all the typical problems of software projects (see Brooks' famous article "No Silver Bullet," which discusses this observation [Brooks87]).

Hamlet and Maybee wonder in their introduction whether software engineering will even attain the status accorded to civil engineering [Hamlet and Maybee01]. Then, Chapter 3 of Hamlet and Maybee's book discusses the question, "Is it really engineering?" They declare that much of the human activity of dealing with objects in the world can be categorized as art, craft, science, and engineering, and wonder where software lies in these categories.

Their analysis of the field of software engineering is based on the examination of the following question: In what ways is software different from other products? Among other arguments, they declare that software is perhaps the most complex human artifact, so there is a broad range for error. A similar message is conveyed by [Sommerville90], who states that a wide range of tools and methods are needed in software engineering because there are no simple solutions to software engineering problems. Indeed, how to select the "best solution" for a software engineering problem is one of the core questions of the discipline. In fact, a more basic question is, what are the attributes of a good software development process and software product? Some examination of these questions is presented in Chapters 2 and 4 in this book.

The preceding opinions about software engineering suggest that recent approaches toward software engineering do conceive of it as a multifaceted discipline. More specifically, it seems that software engineering can be examined from three main perspectives: engineering, scientific, and human. The fact that the human aspect currently gets more attention can be explained by the difficulties involved in managing software projects. The continuation of this chapter presents two frameworks that refer to the human aspect of the field by specifically addressing organizational issues connected to software projects.

TASK

The definitions for software engineering in the previous section illustrate changes in the way the discipline has been conceived of since its establishment in 1968. Chapter 8 presents the history of software engineering, emphasizing the evolution of software development methods. Based on these resources, examine connections between the evolutionary process of software development methods and the changes in the perspectives of software engineering.

The Product versus Process Perspectives of Software Engineering

This section is based on [Floyd87], which contrasts two approaches toward software engineering: the process approach and the product approach. Floyd declares explicitly that she criticizes the product-oriented perspective since it cannot answer questions that are raised from the interaction of software with the human world. This opinion explains clearly why we find it relevant to present in this book the dual perspective suggested by Floyd.

We start by presenting the main characteristics of each approach. According to the product-oriented view, programs are formal mathematical objects, derived by formalized procedures starting from abstract specifications. Accordingly, their correctness is established by mathematical proofs. As can be observed, computer programs are at the center of this approach. In contrast, the process-oriented approach connects the existence of computer programs to people. Specifically, according to the process-oriented perspective, programs are tools or working environments for people. They are designed through learning and communication processes to fit human needs. Their adequacy is achieved in processes of controlled use and subsequent revisions.

The title of Chapter 4 of this book is "Software as a Product." Our intention in titling Chapter 4 in this way is to emphasize the analysis of software systems from the customer's perspective, as a product that should meet customer's needs. The nature of the product-oriented view presented in this section is different. In fact, the nature of the process-oriented view discussed in this section is more similar to the messages delivered in Chapter 4.

TASK

Based on the brief introduction of the process-oriented and the product-oriented approaches:

- Suggest elements of software development environments that can be analyzed from these two perspectives. Analyze these elements from these two approaches.
- Suggest situations in software development that may suffer when different people on the team hold either product-oriented or process-oriented approaches. How can such problems be solved?

The two approaches can be compared on different dimensions. Table 7.1 compares the two perspectives by illustrating their attitude toward three software engineering topics [Floyd87].

The following tasks guide readers in a further examination of the two approaches.

TASKS

1. How does each of the two approaches address the human aspect of software engineering?
2. Discuss connections between the "programs understanding and the role of documents" topic, presented in Table 7.1, and the discussion about pro-

TABLE 7.1 Comparison of the Process-Oriented and the Product-Oriented Perspectives

Dimension	Product-Oriented View	Process-Oriented View
Quality	Quality is associated with features of the product.	Quality is associated with processes of using the product.
Software development	Software development aims at producing one software system.	Software development aims at producing a sequence of related versions of the software system.
Programs understanding and the role of documents	Programs should be understandable only from documents.	Personal discussions and trial uses are indispensable for programs understanding; documents must facilitate these activities.

gram comprehension presented in Chapter 9, "Program Comprehension, Code Inspections, and Refactoring."

3. As part of the process-oriented approach, Floyd recommends writing a project diary and using it as a means to help in the understanding of software systems.
   - Suggest at least five types of items that are worth including in such a diary. Explain how they might contribute to the understanding of software systems.
   - Start writing a project diary for your current software project. Keep reflecting on the writing process (see Chapter 10, "Learning Processes in Software Engineering"). What are your main conclusions?
4. There are additional topics according to which the product-oriented and the process-oriented perspectives can be compared. Floyd, for example, mentions, among other topics, the following: errors, user competence, and methods. Suggest additional topics to which the two approaches can be compared. Perform this comparison.

The Agility Paradigm versus the Heavyweight Approach Toward Software Development

Agile software development methods were mentioned in Chapter 2. Special emphasis is placed in Chapter 2 on eXtreme Programming (XP), one of the agile software development methods. In general, the agile approach toward software development has emerged in the last decade as an answer to the unique and typical problems of software development processes.

Agile software development methods emphasize customers' needs, short releases, and heavy testing all along the development process. Fowler explains the differences between agile methods and heavyweight methods [Fowler02]:

Agile methods claim to be adaptive rather than predictive. Heavy methods tend to try to plan a large part of the software process in great detail for a long span of time. This method works well until things change. Therefore, the nature of such methods is to resist change. The agile methods, however, welcome change. They try to be processes that adapt and thrive on change, even to the point of changing themselves.

Agile methods are people oriented rather than process oriented. The agile methods advocate working with human nature, not against it.

The agile software development methods are based on the following four principles, called the Manifesto for Agile Software Development (*http://agilemanifesto.org*):

- Individuals and interactions over processes and tools
- Working software over comprehensive documentation
- Customer collaboration over contract negotiation
- Responding to change over following a plan

This manifesto was created by 17 software practitioners gathered at Snowbird, Utah, in February 2001. They all agreed to use the term "agile" and call themselves "agilists." The manifesto's principles are based on several agile software methods that have been developed by these professionals in order to cope with the typical problems of software development. Although items to the right of the sentences are not ignored, they are not considered as important as items on the left, or the beginning of the sentences. It is interesting to note that although this approach was instituted only three years ago, it has momentum, and many software houses adopt its orientation in general or a specific agile software development method in particular.

Here are several examples of agile software development methods: XP (described in Chapter 2), SCRUM, Adaptive Software Process, DSDM (Dynamic Systems Development Method), and Feature Driven Development. To illustrate the essence of these methods, here are four out of the nine underlying principles of DSDM: focus is on frequent delivery of products, iterative and incremental development is necessary, testing is integrated throughout the life cycle, and collaboration and cooperation between all stakeholders are essential.

TASKS

1. Select three agile software development methods. Explain how each implements the principles of the Manifesto for Agile Software Development.
2. In what sense are agile software development methods similar? In what sense are they different?
3. In what sense do the agile methods correspond to the process-oriented approach described previously in this chapter?

One idea connects the heavyweight methods with the agile software development methods: the paradigm of software engineering (see Chapter 8), consisting of the activities of specifying, designing, coding, and testing. One of the main differences between the heavyweight and agile methods is the frequency in which these activities are implemented in each approach. Generally speaking, the agile methods tend to intertwine these activities more frequently. This frequent implementation of the paradigm of software development throughout the development process enables the agile methods to be more open to changes.

Additional Approaches

We now briefly describe two additional approaches toward software engineering: analogies to other professions and failure and success of software projects.

Analogies to Other Professions

Software engineering can also be examined by its comparison to other professions. One of the more accepted analogies to software engineering is architecture. This analogy is discussed in Chapter 10 when the reflective practitioner perspective, originally developed by Schön for architects [Schön83], is discussed.

Another source that illustrates the analogy between software engineering and architecture is the Worldwide Institute of Software Architects (*www.wwisa.org/wwisamain/index.htm*). Among other ways it illustrates this analogy, the institute presents eight phases that define the role of the architect/developer in the process of software construction. These phases conceptually follow the phases of building construction and architectural services. It is emphasized that these phases are independent of any particular software development method and apply to all software construction projects. The institute declares that many software professionals are already drawing on the analogy with building construction to describe their processes, since it is a true metaphor and it is understandable to clients.

Needless to say, and as mentioned previously in different places in the book, software engineering has its own unique characteristics. Still, we believe that such analogies may help in understanding subtle aspects of the discipline of software engineering.

TASK

Suggest other professions that may be analogous to software engineering. In what sense are software engineering and these professions analogous? In what sense are these professions different from software engineering?

Failure and Success of Software Projects

Chapter 1, "The Nature of Software Engineering," mentions successes and failures of software projects. Specifically, we cite Kent Beck, the creator of eXtreme Programming, who refers to the conceptual change toward software project success and failure that both developers and customers should adopt when they decide to use XP as the project development method [Beck02]. Indeed, the question of what a successful software project is invites debate. One agreement has been reached, though, among the community of software practitioners: they mostly agree that software projects should meet customers' needs.

To illustrate this point, consider the two software projects described in Table 7.2. For each, decide whether it is a success or a failure.

TABLE 7.2 Two Software Projects: Are They Success or Failure?

Project A	Project B
Software for airlines management:	Software for a medical equipment:
• 64 percent of the developers had more than five years of software development experience.	64 percent of the developers had more than five years of experience in software development.
• Delivered on time.	Time to market: 193 percent (of what was expected).
• Not over budget.	Over budget: 200 percent.
• After delivery: the project did not work as was expected.	After delivery: the project works as was expected.

Discussion

- Define a successful software project.
- Define a failed software project.
- How can the success of a software project be measured?

SUMMARY QUESTIONS

1. Find connections between the two main subjects discussed in this chapter: the process-oriented versus product-oriented perspectives of software engineering and the agile software development methods versus the heavyweight software development methods.
2. Review the different chapters of this book. Identify what specific perspective was emphasized in the presentation of each chapter.

FOR FURTHER REVIEW

The fact that software engineering is accepted differently by different groups of practitioners is illustrated also by the industry perspective versus academia's perspective on what should be taught in software engineering programs. Indeed, many discussions address this issue. This topic is also discussed in Chapter 18, "Remarks about Software Engineering Education." Review the different resources presented in Chapter 18 and suggest topics related to software engineering education with respect to which academia and industry may have different perspectives. Suggest ways these gaps can be bridged.

REFERENCES

[Boehm76] Boehm, B. W., "Software Engineering," *IEEE Transactions on Computers*, (December 1976): pp. 1226–1241.

[Beck02] "Extreme Programming: An Interview with Kent Beck," The Cutter Edge, *www.cutter.com/research/2002/edge020903.htm*, September 3, 2002.

[Brooks75] Brooks, Frederick P., *The Mythical Man-Month: Essays on Software Engineering*, Addison-Wesley, 1975.

[Brooks87] Brooks, Frederick P., "No Silver Bullet—Essence and Accidents of Software Engineering," *Computer Magazine* (April 1987): pp. 10–19.

[Floyd87] Floyd, Christiane, "Outline of a Paradigm Change in Software Engineering." In Bjerknes, P. G., Ehn, M. Kyng (Eds.), *Computers and Democracy—a Scandinavian Challenge*. Avebury, 1987: pp. 191–212.

[Fowler02] Fowler, Martin, "The New Methodology," published in *www.martinfowler.com/articles/newMethodology.html*, 2002.

[Hamlet and Maybee01] Hamlet, Dick, and Maybee, Joe, *The Engineering of Software*. Addison-Wesley, 2001.

[vonMayrhauser90] von Mayrhauser, Anneliese, *Software Engineering—Methods and Management*. Academic Press, Inc., 1990.

[Sommerville90] Sommerville, Ian, *Software Engineering,* Sixth Edition. Addison Wesley, 2001.

[Schön83] Schön, Donald A., *The Reflective Practitioner*. BasicBooks, 1983.

[Zelkowitz78] Zelkowitz, Marvin V., "Perspectives On Software Engineering," *ACM Computing Surveys* 10(2), (1978): pp. 197–216.

8 The History of Software Engineering

In This Chapter

- Introduction
- Objectives
- Study Questions
- The Early Days of Computing
- Information Hiding—The First Budding of Software Development Methods
- Abstraction—Another Part of Methods
- The Beginning of Software Development Methods
- The Customer's Task in the Early Days of Software Development Methods
- Abstraction and Information Hiding Come to the Fore
- Software Development Methods Become Part of the Profession of Software Engineering
- Objects Arise
- Agile Methods Enter the Software Engineering World

INTRODUCTION

This chapter presents the history of software engineering. It seems that this history is a combination of three evolutionary processes that occurred in parallel: the history of computers, the history of programming languages, and the history of software engineering methods.

Among the many interesting observations that can be discussed with respect to these historical processes, we emphasize three aspects. The first is that although the history of these areas is short (about half of a century), it went through so many events that each of these events influenced the future in a unique way. The second is that the three types of histories are intertwined and none can be isolated. The third is that human aspects have a significant role in these historical processes. That is, we cannot describe these processes without examining how people behaved and reacted to what happened around them. Thus, this chapter is presented like a story, with heroes, their behaviors, and details that make the stories vivid.

OBJECTIVES

- Readers will become familiar with the main stages of the history of software engineering.
- Readers will become familiar with the nature of the historical development of software engineering.
- Readers will gain a historical perspective on the topics discussed previously in the book.

STUDY QUESTIONS

1. What is the software engineering paradigm?
2. What led to the development of the concepts of abstraction, information hiding, and iteration in software development processes?

THE EARLY DAYS OF COMPUTING

At the very beginning of stored programs, and thus software, computer hardware was relatively large. Nevertheless, only one person could use a computer at a time, making a computer the biggest personal computer ever.

There were several hundred years of frozen-programmed or hard-to-program mechanical analog computers, most in the 1600 to 1930s period, before the early electronic digital computers. As an example, Dr. Helmut Hoelzer designed and produced a complicated analog computer to help guide the A-4 (or V-2) rocket [Tomayko85]. However, due to the specificity of electronic analog computing, the actual guidance system could not be used for other rockets, although the principles are the same. Similarly, the analog computers used for the flight control system on the General Dynamics F-16A and B were not transferable exactly to the Lockheed F-117A, but the latter airplane's flight control system used the same analog types of circuits in a different arrangement (later, a version of Hoelzer's computer was transferred to several U.S. launch vehicles) (see Figure 8.1).

FIGURE 8.1 A Redstone rocket lifts off (using guidance vanes in the exhaust). © Redstone Arsenal Historical Information

Later electronic digital computers could be operated via software. Most, like the Electronic Numeric Integrator and Calculator (ENIAC), were still difficult to program or could hold only one procedure at a time. While University of Pennsylvania faculty, like John Mauchly, and graduate students, like J. Presper Eckert, built the ENIAC, meetings were held between them and the famous mathematician John von Neumann to look ahead. One thing that was immediately obvious to them was that without the ability to easily change the program, electronic computers would be nearly useless. At one of these discussions, the participants realized that programs and data could be represented using the same techniques used in computers.

Von Neumann wrote, in 1946, "A Draft Report on the EDVAC," which, unlike the ENIAC, used binary representation of numbers (the ENIAC was a decimal machine). The "stored program concept" described in this paper was derived by the group at Penn collectively.

Eckert and Mauchly left Penn to form a company that eventually built the Universal Arithmetic Calculator (UNIVAC). The EDVAC design formed the core of the "Johnniac," named after von Neumann, at Princeton University's Institute for Advanced Study. Meanwhile, the idea crossed the ocean, coming to fruition first in England. Manchester University built the little-known "Baby" computer with stored program capability in 1948. Maurice Wilkes of Cambridge University's Mathematical Laboratory built the much more capable, and better known, Electronic Delay Storage Automatic Computer (EDSAC) in 1949.

The EDSAC's memory, a big-for-the-time 512 words of binary numbers, was *acoustic*. This was accomplished with glass tubes mostly filled with liquid (room temperature) mercury. One end of the glass tube had a speaker of sorts, and the other end had an amplifier. Ones were stored as high amplitude sound waves in the mercury. Zeroes were low-amplitude waves. When ones or zeroes reached the amplified end, they would be strengthened and returned to the other end (see Figure 8.2). This is called a *Delay Line*, since a number took some time to be read. When one number needed to be read, a counter circuit would count the numbers as they went by in a specified bank. Eckert devised a similar memory scheme during World War II.

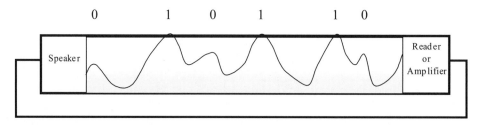

FIGURE 8.2 A Mercury Delay Line memory.

TASKS

1. What is the "stored program concept?" What is the important insight in this observation?
2. Find out what a Williams Tube memory is. Discuss whether it is a better technology than the Mercury Delay Line.

INFORMATION HIDING—THE FIRST BUDDING OF SOFTWARE DEVELOPMENT METHODS

During the few years of programmable computers until the early 1950s, no software development methods existed. It can be argued that small memories made methods unnecessary. *Abstraction* and *information hiding* became the basis of all future methods.

Computer scientists took another 15 years to recognize that abstraction was needed. Information hiding was used long before it was formally recognized. David Parnas, then teaching at Carnegie-Mellon, where he had graduated, wrote the famous "On Decomposing Programs Into Modules" article that showed how information hiding worked [Parnas72]. Fundamentally, Parnas said that modules have "secrets." They do things, but we do not know how. We know just the interfaces and names of the procedures.

For example, let us consider a linked list using pointers. FORTRAN has no pointers in early versions. We choose to implement linked lists in FORTRAN using an array. If other programmers on a project know this, they can find the value of some item in the list by array access. This would defeat the purpose of information hiding.

The EDSAC used subroutines. It was fed input on paper tape. One enterprising programmer, rushing upstairs to use the EDSAC, trailing the paper tape, realized that the first few instructions and several print routines were identical to ones others made. The programmer borrowed somebody else's tape of those. Eventually, these partial tapes migrated to hang on the wall near the operator, captured in many photos. As programmers came in, they would grab the setup tapes and print tapes to read in as needed. Eventually, other routines became standard, mostly mathematical terms such as *square root* or *differentiation*. These were added to the collection near the operator. Thus, information hiding was off to a start, although "secrets" were not kept reliably.

ABSTRACTION—ANOTHER PART OF METHODS

Abstraction had a longer genesis. Throughout the 1950s, while subroutines came to be better known, the concept of abstraction was rarely used. There were large systems built in the Western Hemisphere at this time, such as the air defense program Semi-Automatic Ground Environment (SAGE). Eventually, in the early 1960s, computer scientists started to discuss "stepwise refinement" as a way to reduce complexity.

This was the origin of abstraction. The highest level of a program's development is also its highest level of abstraction. As the program is stepwise refined, the abstractions are closer to reality. As an example, a program may "add two numbers." At the next level, two numbers are input, added together, and one is output. Therefore, there are three parts to that abstraction. If we work just on the input part, we can establish a range, verify that range, and so on. If we are doing the adder, we can only worry about how two numbers get to the next abstraction. We do not have to worry about establishing the range, testing the numbers, or any messages. The developer of that abstraction can hide that information from us.

TASK

Look at several textbooks that deal with programming and other books about abstract data types. How do they define abstraction? In what sense is abstraction important in computer science in general?

In the early 1970s, both of these concepts could be formalized. Parnas wrote his article. Barbara Liskov of the Massachusetts Institute of Technology (MIT) first drew attention to abstraction in a formal way [Liskov74].

At about this time, the term *software engineering* became prevalent. It seems to have first been used by the organizers of the 1968 North Atlantic Treaty Organization software conference in Garmisch, West Germany to spark debate [Naur and Randell69]. There is evidence that Douglas Ross used it earlier in a course at MIT [Ross88]. It is ironic to read the proceedings of this conference nearly four decades later and realize that most attendees went to learn other people's solutions, only to find that everyone had the same problems.

More about abstraction, and especially on its practical aspects, can be found in Chapter 11, "Abstraction and Other Heuristics of Software Development."

THE BEGINNING OF SOFTWARE DEVELOPMENT METHODS

During the 1950s and 1960s, some programmers used abstraction and information hiding inadvertently in their programs. The important thing is that by 1975, both were better known. Moreover, engineers used the initial software development method, or, in this case, life cycle, while working in a form of obscurity. Winston Royce wrote about this method in 1970, in an article entitled "Managing the Development of Large Systems: Concepts and Techniques" [Royce70]. What he describes in that article looks like the "documentation life cycle," because each milestone is based on releasing some document. It is now called the "Waterfall Life Cycle," because it looks like Figure 8.3.

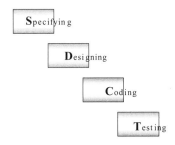

FIGURE 8.3 The Waterfall Life Cycle.

Note that the Waterfall model contains the terms *specifying*, *designing*, *coding*, and *testing*, as we will meet them again. It also flows from top to bottom; hence, the waterfall metaphor. As a natural waterfall does not flow uphill, this kind of waterfall implies that we are done with an activity before we proceed. This means that we are done specifying before designing begins. Those who expected that requirements changes ended soon after being specified were terribly wrong. They experienced the waterfall as shown in Figure 8.4.

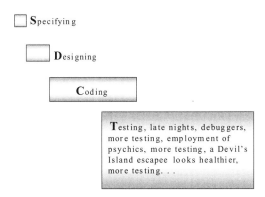

FIGURE 8.4 How the Waterfall Software Life Cycle often looks.

132 Human Aspects of Software Engineering

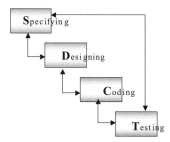

FIGURE 8.5 The Waterfall Software Life Cycle with feedback.

Soon, the waterfall was modified by feedback loop, thus Figure 8.5.

This feedback loop enables engineers to react to change. If sufficient changes happen to the requirements, then the design can be changed. If coding reveals a design change, then it can be propagated backwards. This proved to be too cumbersome or ineffective, due to changes performed in things that were already considered done. Therefore, another change to the waterfall became prevalent, the use of prototypes (see Figure 8.6).

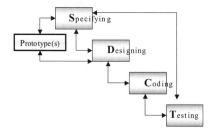

FIGURE 8.6 Prototypes in the Waterfall Software Life Cycle.

Discussion

Describe in detail the differences between the last three models. What do these differences imply with respect to software developers?

THE CUSTOMER'S TASK IN THE EARLY DAYS OF SOFTWARE DEVELOPMENT METHODS

In the beginning, the clients' role was to examine prototypes. That is, clients did not take part in requirements formulation. There turned out to be two kinds of prototypes. One demonstrated the interface to the users. This placated the IKIWISI ("I'll Know It When I See It") users. The other tried out different things in the requirements to check feasibility.

The problem with the former type of prototype is that the closer a developer gets to a sensible interface, the more the client wants it. As tools like Visual Basic™ come to dominate the user interface market, the interfaces appear like "real" software. Clients who do not know the purpose of interface prototypes want to take the "product" with them, not realizing that there is little behind the curtain. This is why we often tell our developers to prototype the user interface using Post-It™ notes on a whiteboard. We have never seen clients leave a room with a whiteboard under their arms! It is obviously not necessary to code the interfaces, unless there is doubt that the physics prototypes can be done. Just do not show clients the coded version—they'll often take it with them, expecting it to be a product.

The second type of prototype, the feasibility study, has its own possible problems. Some years ago, one author (Jim Tomayko) led the production of a museum exhibit. The exhibit was a kiosk that allowed patrons to try the counterintuitive task of orbital rendezvous. As with many museums, this one heavily depended on its patrons for support, one of which had donated an early model of the Apple, Inc., Macintosh™. The curator was especially concerned about accuracy, so one question became whether the small computer could in reality do the mathematics of orbital rendezvous while displaying the Space Shuttle orbit and a hypothetical space station as a rendezvous target.

One subteam was dispatched to do the interface, while a single mathematically oriented (we thought) developer did a prototype of the orbital calculations. Both worked well, and both worked tightly within time ranges, so we proceeded with development. During a design review, it came out that the feasibility prototype we had used for the orbital calculations was flawed. The prototype had adjusted the gravitational constant (well, it was *supposed* to be constant) G until the equations worked. The equations, although altered for "accuracy," still worked simultaneously with the interface, fortunately.

The use of prototypes in industry encountered significant resistance. This resistance peaked when some managers saw prototype code being thrown away. That struck the wrong chord with those spending money on something that they thought was concrete. The idea of having the specification and design explored and improved up front eventually became attractive.

ABSTRACTION AND INFORMATION HIDING COME TO THE FORE

Therefore, methods during the 1970s and early 1980s were based on the prototyping version of the Waterfall Software Life Cycle. Abstraction and information hiding came to the forefront. So much so, in fact, that the U.S. government sought to devise a new language that embodied these two concepts from the start. The languages of the day, such as Pascal, could be manipulated into information hiding, since the idea had been around for so long and had penetrated language designers' conception of programming languages. Except for a couple of academic experiments, abstraction *per se* was not institutionalized.

Discussion

How can the last described phenomenon be explained?

The language Ada rectified the situation [Ada83]. The language was named after Ada Augusta Lovelace, daughter of Lord Byron, who was the first programmer of Charles Babbage's mechanical computer—the Analytical Engine. The language had all the features of a third-generation (beyond machine code and COBOL//FORTRAN) language as well as two parts to every procedure and package: a *specification* part and the *body*. They were separately compliable. The keyword *private* in the interface meant just that. It did not matter how the body was built; the programmer did not need to know. As an example of how this aids both abstraction and information hiding, let us consider a stack.

A stack has three operations, *push*, *pop*, and *empty*. Hopefully, the last procedure is done first so we can see if *pop* is possible to prevent trying to pop an empty stack (think of a stack of cafeteria trays). These three operations are defined as part of the specification part. If it is an Ada generic package, we do not even need to assign a type. Later, when we include the package into our assembly of parts, we can assign this generic to be character or integer type, for example, meaning that this can be a character or integer stack. The body can be written at any time.

TASK

How does such a feature of a programming language influence the programming task from the developer's perspective?

The specification part, thus, is an example of abstraction, since it represents operations as an abstract stack. The body contains the secret of whether the stack is a linked list or an array, perhaps. Therefore, information hiding is present and a programmer cannot cheat and derive a data item by indexing the hidden structure.

TASKS

1. Suggest additional examples of such implementation of the idea of abstraction.
2. Read about data abstraction and procedural abstraction. What is common to these two types of abstraction (in other words, what is the abstract idea behind them?—pun intended). In what case is each used?

Ada, the language, went through several iterations of requirements and design before release of the first version in 1983. The U.S. government thought that Ada would replace hundreds of languages in use and it was prescribed for government contractors soon after. However, within a decade or so, even after being fully converted to object orientation in Ada95, extensive waivers and natural chafing against a mandated requirement killed it. The government ensured compliance by granting certifications to compilers that passed a test suite they maintained. Although several major military and space projects use Ada, it never caught on. The chief argument is that it is too big, a "language for all seasons."

SOFTWARE DEVELOPMENT METHODS BECOME PART OF THE PROFESSION OF SOFTWARE ENGINEERING

At around the turn of the 1970s to 1980s, two offshoots of the waterfall model became evident in software development. Both answered business's call to be "faster." Some companies using the waterfall model took shortcuts to achieve speed. Since lack of speed became apparent toward the end of the life cycle as the projected shipping date approached, often the cuts came in testing.

TASK

Look at Chapter 2, "Software Engineering Methods." How do software development methods overcome this last described phenomenon?

One way of increasing the speed of development was to use prototypes as the basis of development (see Figure 8.7).

In this variant, a prototype could be buttressed with additional code and documentation to make it releasable, but this was the first code cut if a budget crunch appeared, leaving only the prototype, with bare documentation.

Another variant on the waterfall model, which actually supported it and found its way into many methods, is the Iterative Model (see Figure 8.8).

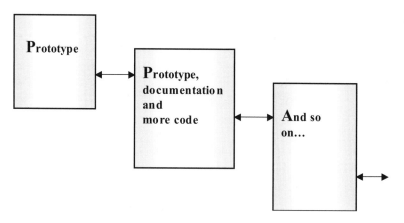

FIGURE 8.7 The Evolutionary Development Model.

The idea here is that we know what we want to produce, but we plan several iterations of work, each saleable. The problem was the defect spike after the first and other releases, necessitating a schedule slip on Release Two. Each subsequent release also slipped. The added time of these slips could lead to a release that was no longer valuable. However, early value was the main concern, and it would hold up for a few releases.

The emphasis in these later two versions of the Waterfall Software Life Cycle was on repetition. As the 1980s wore on and dissatisfaction with the Waterfall Software Life Cycle, on which most methods were based, came to a head, a new life cycle design came from a trusted source. The trusted source was Barry Boehm [Boehm81]. The new life cycle was the Spiral Model (see Figure 8.9).

The Spiral Model came from a desire to reduce risks, but it was iterative in style (for additional information on the use of this model see Chapter 2). The first pass through specifying, designing, coding, and testing is for a component of the software

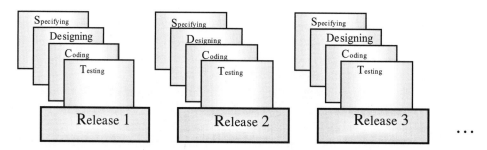

FIGURE 8.8 The Iterative Development Model.

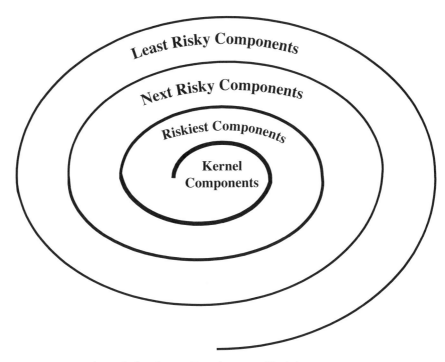

FIGURE 8.9 The Spiral Software Development Model.

that we knew least about. Therefore, its configuration, elapsed time for development, and cost are unknown. This component represents the greatest risk (if we know how to do something, it is easy), and it would be completed early in the process! Each spiral has the same activities concentrated on items of increasingly lesser risk.

Task

Describe an evolutionary process behind the Spiral Model.

Steve McConnell immortalized this combination of risk reduction and iteration as a "best practice" in his "Rapid Application Development" [McConnell96].

Notice that all the figures traverse the activities of specifying, designing, coding, and testing. Some do this several times. This is because these activities have become the *paradigm* of software development. We use the concept of paradigm in the way Thomas Kuhn meant it originally in his book [Kuhn62]; that is, not as *a* way of doing things, but *the* way. It is more or less true that these four activities, plus the concepts of abstraction and information hiding, served as the basis for iterative methods. When the majority of developers have accepted these four activities, they become a paradigm.

OBJECTS ARISE

In the 1980s and 1990s, "objects" came to the fore. If you look at the way objects were originally derived, by identifying proper nouns, common nouns, and verbs, you can see that they are directly linked to the simplest parts of natural language. Objects and classes embody abstraction and information hiding at the most direct level. It is easy to think of them as commonplace things, far from the difficulty of most of software engineering.

Grady Booch in his second book on Ada described how to derive objects from natural language, using common nouns as objects or classes of objects, proper nouns as instances of objects, and verbs as interfaces [Booch94]. Later, he wrote a series of books on object-orientation [Booch86].

The company he worked for, Rational, began as a maker of machines using Ada as native code, thus making the boxcar of a language very fast. His two Ada books fit the company's desire to have companion explanations of doing software. Booch moved into objects as Rational decided to reposition itself from a hardware company to a software tool company. They also acquired as employees Ivar Jacobsen and James Rumbaugh. Rumbaugh was the Object Modeling Technique (OMT) originator.

These two, with Booch, became known as the "Three Amigos" in the community. They developed a method for objects that use all the previously described activities and concepts. This method is the Unified Process (UP), and it uses the tools of the Unified Modeling Language (UML). Since Rational's UML tool suite, Rose,™ has become so identified with the Three Amigos, and has sold so well, UP is better known as RUP (Rational Unified Process). Thus, UP takes what engineers have learned about "growing" software through iterative life cycles and couples it with abstraction and information hiding [Jacobsen99]. There is more about the UP and UML in Chapter 2.

UP is large, UML itself is quite large, and documentation is a big part of the method. The way we look at it, documentation that eases the repair or transition processes is all right, but sometimes UP documents put people off. Excessive documentation is partially solved by the agile methods, as is described in what follows.

AGILE METHODS ENTER THE SOFTWARE ENGINEERING WORLD

As described in Chapter 2, there is a newer method, eXtreme Programming (XP), one of the "agile" type. XP implements the value of communication by open-space development areas and customers on site. This way, if either the client or developer has a question, it can be addressed directly and simply. One idea is that most of the questions are asked during development, so little documentation is needed. In an atmosphere of seemingly constant change, XP is based on the proclamation that anything written down ceases to be current the instant it is published. So, why waste time?

Other values in XP are simplicity, feedback, and courage. Practices are chosen to implement these values. Chapter 2 describes these practices emphasizing their human aspects.

Since XP and UP are described elsewhere ([Beck00] and [Jacobsen99], respectively), and in part in Chapter 2 of this book, we will not spend time on them here.

TASKS

1. Identify evolutionary processes in the development of software development methods.
2. How is it that agile approaches were not "invented" before?

FOR FURTHER REVIEW

Look at the Web for the concept "evolution of programming languages." Identify the main steps of this evolutionary process. In your opinion, what are the main stages of this evolution? Describe how each step of this evolution answers problems of its previous step.

REFERENCES

[Ada83] Ada Joint Program Office, *Military Standard Ada Programming Language*, Washington, DC, 1983.

[Beck 00] Beck, Kent, *Extreme Programming Explained*, Addison-Wesley, 2000.

[Boehm81] Boehm, Barry W., *Software Engineering Economics*, Prentice Hall PTR, Englewood Cliffs, NJ, 1981.

[Booch94] Booch, Grady, *Object Oriented Analysis and Design*, Second Edition, Redwood City, CA, 1994.

[Booch86] Booch, Grady, *Software Engineering With Ada*, Second Edition, Menlo Park, CA, 1986.

[Kuhn62] Kuhn, Thomas, *The Structure of Scientific Revolutions*, Chicago, 1962.

[Jacobsen99] Jacobsen, Ivar et al., *The Unified Software Development Process*, Addison-Wesley, Boston, 1999.

[Liskov74] Liskov, Barbara H., and Zilles, S., "Programming with Abstract Data Types," *SIGPLAN Notices* (April 1974): pp. 50–59.

[McConnell96] McConnell, Steve, *Rapid Development*, Microsoft Press 1996.

[Tomayko85] Tomayko, James E., "Helmut Hoelzer's Fully Electronic Analong Computer," *Annals of the History of Computing,* (July-September 1985): pp. 227–240.

[Royce70] Royce, Winston W., "Managing the Development of Large Software Systems: Concepts and Techniques," Proceedings, WESCON, August, 1970.

Part III

Software-Human Interaction

Chapter 9, Program Comprehension, Code Inspections, and Refactoring

Chapter 10, Learning Processes in Software Engineering

Chapter 11, Abstraction and Other Heuristics of Software Development

Chapter 12, The Characteristics of Software and the Human Aspects of Software Engineering

9 Program Comprehension, Code Inspections, and Refactoring

In This Chapter

- Introduction
- Objectives
- Relevance for Software Engineering
- Program Comprehension
- Theories of Program Comprehension
- Code Inspections (Code Review)
- Refactoring
- Summary Questions

INTRODUCTION

This chapter discusses three closely related topics. The first is program comprehension, which is about how software developers understand computer programs. The second is code inspections (or, as it is sometimes called, code review), which is a team activity in which the team members, together, review a piece of code, to find mistakes and suggest improvements. The third topic is refactoring, which is about reshaping the code to improve its readability and comprehension without changing its external (observable) behavior.

Interconnections among the three topics are clear: when the team reviews the code and suggests ways to improve it, all team members improve their understanding of the code; when the code is refactored, its readability is improved, and as a result, it becomes more comprehensible. In the discussion that follows, computer programs are approached from three perspectives that support each other: a cognitive perspective (program comprehension), a social perspective (code inspections), and a technical perspective (refactoring).

OBJECTIVES

- Readers will acknowledge the importance of programming style and its influence on program comprehension.
- Readers will observe connections between programming style and the daily life of software developers.
- Readers will understand the benefits of code inspections and refactoring.
- Readers will increase their awareness of situations in which code inspections and refactoring processes should be integrated.

RELEVANCE FOR SOFTWARE ENGINEERING

Software engineers produce code (no matter which development method they choose to work in). This production includes the writing and reading of computer programs. On the surface, it seems that during the writing process one can ignore the others who participate in the development process. However, because software is an evolving creation, and the development of any piece of code is carried out (in most cases) by more than one programmer, it is more reasonable to assume that the code one writes will be read and will have to be comprehensible by others at some stage in the future. Thus, one cannot ignore the perspective of others during the processes of code writing.

This chapter focuses on considerations a software developer should address in order to produce comprehensible code. Because it is not a trivial task to produce qualified and comprehensible code at first writing, some reworking, reshaping, and redesign processes are required. Two techniques that may help in the production of more comprehensible code are described in this chapter as well: code inspections and refactoring.

Code inspection is usually carried out by a software team that reviews the code, suggests improvements, and helps find mistakes; refactoring is done by the individual (or pair of programmers) who develops the code. It guides developers to improve the code structure and shape during and after the code production. Refactoring improves code readability and understandability and, consequently, makes it cheaper to perform future modification on the code.

The relevance of each of these topics to software engineering is clear. Think about a situation in which you develop a piece of code and it seems to work very well according to all the defined tests. Since you are eager to continue working on your development tasks, you may conclude that you may move on to your next task. This would be true if you lived in an ideal world in which any piece of tested code is considered a finished product (that will never be altered) as soon as it passes all tests. However, in our software world, this is not the case, and it seems more logical to assume that this piece of code will be altered (or at least will be viewed) in the future for different reasons. Thus, as much as you dislike it, before moving on to the next development task, you will have to invest more time, work, and effort to improve the code readability and comprehensibility.

Code inspections and refactoring may help you in overcoming the natural tendency to move on before your code is polished. The message that this chapter conveys is that although program comprehension is about the code, it has also a lot to do with people's thinking, attitudes, experiences, and even emotions.

PROGRAM COMPREHENSION

Study Questions

1. In your opinion, what is program comprehension? In other words, what does it mean to comprehend a computer program?
2. In what activities during the process of software development is program comprehension relevant? Important?
3. Recall five situations in which you failed to understand a simple computer program. Can you explain the reason for these difficulties?
4. Recall situations in which you succeeded in reading and understanding big computer programs easily. What characterized these computer programs?
5. Develop two computer programs that perform the same task. One program is easily comprehensible; the second program is difficult to comprehend. What guided you in the development of each program?
6. In your opinion, what are the five most important factors that influence the comprehensibility of computer programs?
7. Suggest five heuristics that may help software developers in the process of comprehending computer programs. Refer to rules that can be used during the writing process and to guidelines that can be applied during the comprehension process.

The topic of program comprehension gets a lot of attention. This fact is reflected, for example, by the rich *Annotated Bibliography about Code Reading and Program Comprehension*.[1] A closer look at the annotated bibliography reveals that the topic is complex and has many subtopics, such as problems in program comprehension, factors that influence the comprehension of computer programs, relationships between professional experience and program comprehension abilities, heuristics to be applied in the process of program comprehension, and pedagogical attempts to help students gain experience in program comprehension. It is clear that program comprehension influences directly the daily life of software developers. For example, based on four experiments that demonstrate the influence of typographic style (source code formatting and commenting) on program comprehension, Oman [Oman90] concludes that about one-half of the maintenance programmer's time is spent studying the source code and related documentation. Fjeldstad and Hamlen [Fjeldstad and Hamlen83] observe that during enhancement or maintenance process of software systems, programmers study the original program about 3.5 times as long as they study the documentation of the program, and just as long as they spend implementing the enhancement.

One basic and interesting question to be posed is, why is program comprehension discussed at all? Can't we read code as we read other written documents, such as books, letters, program documentation, and so forth? The answer, of course, is "No." Computer programs are expressed in a unique way that requires different reading skills than the skills needed to read and comprehend other documents.

TASKS

1. Identify several features/characteristics of computer programs that make their understanding different from the understanding of other written documents.
2. Are the characteristics you listed in Question 1 dependent on the programming language in which the computer program is written?
    2a. If yes, in what sense? Suggest situations in which there is a relationship between the comprehension of computer programs and the programming language by which they are expressed.
    2b. If no, what characteristics of computer programs make their comprehension independent on the programming languages in which they are written?

Theories of Program Comprehension

A comprehensive article in which Davies [Davies93] outlines the main research studies about program comprehension concludes that although a wide range of perspectives have been adopted by studies that address the strategic aspects of programming skills, no clear picture emerges from this work. Among other implications, this conclusion says that program comprehension is still in search of one coherent theoretical framework. Because comprehension is so relevant to the daily life of software developers, the lack of a framework raises some practical questions, such as: How should software developers cope with program comprehension? How does the fact that no clear picture with respect to program comprehension has emerged so far influence code quality and development environments?

Davies [Davies93] suggests that multiple knowledge sources should be used in order to understand the behavior of computer programs. This perspective will guide the continuation of this section about theories of program comprehension. Specifically, we address factors that influence the process of program comprehension and strategies and actions adopted by software developers when they cope with the comprehension of computer programs.

We start by examining what it means to understand a computer program. Is it limited to understanding all the variables? All the objects? All the data structures? Knowing what all the procedures are? The role of all functions? Is it important to understand the overall behavior of the program or only the behavior of specific parts?

According to Biggerstaff, Mitbander, and Webster [Biggerstaff94], one understands a computer program when one is able to explain different aspects of the program (such as its structure and its behavior) in a way that is qualitatively different from the way by which these aspects are expressed by the source code (p. 72). According to Littman, Pinto, Letovsky, and Soloway [Littman87], to understand a program, a programmer must be familiar with the *objects* that the program manipulates and the *actions* that the program performs.

TASK

Select the longest computer program that you developed during the last year. According to the preceding definitions, explain what it means to comprehend that particular program.

Based on [Fjeldstad and Hamlen83] and [Brooks83], we present a list of factors affecting program readability and comprehension:

> **Representational factors and intrinsic properties of the program text:** This includes factors such as the text's length, names of variables, procedure and other identifiers, the programming language, the nature and inclusion of comments, indentation, module structure, and typographic factors such as upper- and lowercase fonts.
>
> **Type of computation the program performs:** The difficulty of comprehending a computer program depends in part on the difficulty of understanding the problem the computer program solves. This happens also in cases when the complexity of the problem is different from the complexity of the resulting program.
>
> **Reason the task is performed (such as modification, debugging):** Different tasks require varying levels of comprehension of different aspects of the program. For

example, a programmer whose task is to modify the output format will be more concerned with the output statements in the program and less concerned with the major control structure than a programmer who attempts to find a bug that causes the program to produce wrong values. The task also affects the way the programmer searches the code listing (local view vs. global view).

Differences among the individuals performing the task: This includes factors such as the programmer's general knowledge, programmer skills and professional experience, and comprehension strategy applied (for example, by hierarchy or by input-output functions).

TASK

Select five of the preceding factors and exemplify them with the program you worked on in the previous task. In that particular program, do these factors support the comprehension of the program or do they disturb its comprehension?

Littman, Pinto, Letovsky, and Soloway [Littman87] collected data from experienced programmers as they enhanced a personnel database program. Based on the data analysis, two strategies for program understanding were identified:

The as-needed strategy: When one uses this strategy, the focus is on the local behavior of the computer program. As a result, since the programmer does not approach the modification task with a wide understanding of the program, it is necessary to gather additional information about the program structure and the way it operates during the program modification.

The systematic strategy: In this case, the target is to understand the global behavior of the computer program. Accordingly, the programmer wants to understand how the program behaves before the modifications are made. For this aim, the developer gathers information about the causal interactions of the program's functional components. Such knowledge permits the programmers to design modifications that take these interactions into account. In most cases, this strategy leads to successful program modification.

TASK

In what cases does it make sense to work according to the as-needed strategy? In what cases does it make sense to work according to the systematic strategy? For each case, explain why it fits the specific strategy.

Chapter 10, "Learning Processes in Software Engineering," deals with just that. In the *Learning Organizations* section of Chapter 10, we discuss the concept of mental models. Here we refer to mental models as the programmers' conception of a computer program. With respect to the task of program comprehension, the following types of mental models have been identified [Littman87]:

Weak mental models of a program: Mental models that contain only static knowledge about the program.

Strong mental models of a program: Mental models that contain both static knowledge about the program and causal knowledge about its behavior.

TASK

Suggest connections between the strategies used for program comprehension (the as-needed strategy/the systematic strategy) and the mental model of a program programmers construct in their mind (weak/strong mental models).

In the rest of this section, we focus on actions that programmers perform in the process of program comprehension. Specifically, Vans, von Mayrhauser and Somlo [Vans99], and von Mayrhauser and Vans [vonMayrhauser93] address the topic of program comprehension through the lens of abstraction (see Chapter 8, "The History of Software Engineering," and Chapter 11, "Abstraction and Other Heuristics of Software Development"). They describe actions, performed by programmers during software maintenance, on the following three levels of abstraction: *Program* level (code), *Situation* level (algorithmic), and *Domain* level (application). By levels of abstraction, we refer to the level of details and the type of inquiries one deals with. It is clear that during a program comprehension task, programmers should move between levels of abstraction according to the type of knowledge they look for and the actual action they perform.

TASK

In your opinion, what types of actions are characterized by each level of abstraction?

In what follows, we present several examples of actions performed at each level of abstraction. We start by examining the high level of abstraction. Full lists of actions for each level can be found in [Vans99].

- Domain level (application)
  - Actions and comments that indicate that the programmer seeks high-level understanding. For example: "I'm thinking how this class contributes to the solution of the problem."

- Comments that refer to high-level understanding strategy. For example: "First I will read the names of the classes and the methods, then. . . ."

TASK

Why do these actions reflect thinking that is characterized by a high level of abstraction?

- Situation level (algorithmic)
  - Questions or inquiries that refer to the application area or situation. For example: "I wonder if they all used the same coding standards."
  - Comments that describe how the programmer will or would have approached understanding. For example: "First I need to understand why and how this method performs this complicated calculation, then. . . ."
- Program level (code)
  - Statements that are word-for-word reading of comments in the code. For example, "The comment declares explicitly that. . . ."
  - Comments that describe a change in the code when it is carried out.

TASKS

1. For each of the actions listed in the preceding section, add at least two phrases that illustrate them.
2. Recall whether you have ever used statements like the ones mentioned or others that belong to these three categories. Describe these situations.
3. Select three of the actions listed in the previous section and describe situations for which the selected actions are useful.
4. Find source code that you or your colleagues developed at least three months ago. Identify one possible upgrade and perform it. Obviously, during this process you will have to comprehend the code. Reflect on the program comprehension process you went through. Identify what strategies you used. If appropriate, suggest ways to improve the code to make its comprehension easier.
5. Ask a friend to comprehend a program that you understand very well. Encourage your friend to think aloud and tape-record what is said during that process. Transcribe the monologue and identify what heuristics of those mentioned previously were used during the process of program comprehension and how your friend moved between the levels of abstraction.

Beyond the classification of the activities that software developers carry out in their attempts to comprehend computer programs, the aforementioned researchers looked at thinking patterns during the process of program comprehension. Among other interesting results, they present the following:

- Programmers with little experience in the domain work at lower levels of abstraction until enough domain experience allows them to make connections from the code to higher levels of abstraction.
- Programmers with domain experience but little knowledge about the software also work at lowers levels of abstraction, but use their knowledge of the domain to make direct connections into higher levels of abstraction of the program model.
- Code size affects the level of abstraction on which maintenance engineers concentrate while building a mental representation of the program. As the code size increases, developers work less with low-level program details.

TASK

Explain the preceding patterns. How do they support the process of program comprehension?

Activity

Phase 1: Split into pairs. Each pair constructs a computer program that has to be as comprehensible as possible. During this attempt to write a comprehensible computer program, reflect on the guidelines you follow.

Phase 2: Switch pairs. Each pair comprehends the code that was written by another pair. During this process, reflect on what heuristics you use in the process of program comprehension.

Discussion

1. Is there a connection between the software development method (see Chapter 2, "Software Engineering Methods") by which a computer program is developed and its comprehension process?
2. In what ways should a general software development method address another of program comprehension? In other words, if you have to design a software development method, would you address the topic of program comprehension? If yes, how? If no, explain why.

CODE INSPECTIONS (CODE REVIEW)

Study Questions

1. What is code inspection?
2. What is the purpose of code inspection? What benefits might a software team gain from reviewing the code it develops?

3. How are code inspections connected to team learning (see Chapter 10)?
4. How is code inspection connected to program comprehension?
5. How does code inspection differ from a review carried out by the individual (or pair) who wrote the code?
6. Suggest three reasons a software team would avoid code inspections.
7. Suggest three mechanisms/procedures that may ensure that software teams review their code on a regular basis.
8. There are companies that offer code inspection services. Find three examples of such companies on the Web. What kinds of services do they offer? What needs of other software houses do these code inspection companies fulfill? Why would a software house hire such a service instead of reviewing the code internally?
9. Read the article "Group to Boost Code Review for Linux" at *http://zdnet.com.com/2100-1104-830255.html*. How does the fact that Linux is Open Source software change its code inspection process?

This section focuses on code inspections and examines connections of code review to program comprehension and refactoring. All these processes are conducted in order to produce code of high quality.

In general, code inspection processes are exploration activities, conducted by the entire team, in order to recognize defects in the code and improve its readability. Inspection has several advantages over other scanning code activities.

First, it helps catch problems of different kinds in the code earlier in the development process. Here are three topics, out of many possible topics, that are important to find out early in the development process and that may come up during a code inspection process: the code does not do what it aims to do; the code is not readable (and some refactoring should be done to improve its readability); and there are better algorithms to implement the desired behavior. No matter what comes up from these options, some code improvements should be made.

Second, on the individual level, code inspection is an opportunity for everyone to be updated with what the others on the team develop, what has been accomplished so far, what strategies were adopted by other team members, and what common problems they faced. In addition, novices may learn from experienced team members.

Third, on the organizational level, because code inspection keeps knowledge within the company and leads to knowledge sharing, it supports the management of company knowledge (see Chapter 10).

TASK

Here are several questions to be asked during a code review process, suggested by Macadamian.[2]

> **General code smoke test:** Does the code execute as expected? Do you understand the code you are reviewing? Has the developer tested the code?
>
> **Comments and coding conventions:** Does the code respect the project coding standards? Are complex algorithms and code optimizations properly commented?

These questions are only a few examples taken from a suggested long list of topics to be addressed and questions to be asked. Because there are so many topics to think about, how can a code-inspection process be managed? How can we ensure that we do not skip any topic? Suggest a procedure for managing a code inspection process so that the important issues are not abandoned.

It is clear that code inspection supports program improvement and hence comprehension. This is achieved mainly by the feedback that all programmers get from the other team members during the code inspections. This feedback can refer to different aspects that may influence the code readability and comprehension. In the next subsection, we focus on refactoring—the actual improving of the code. With respect to refactoring, we also discuss reasons that explain why and when programmers may avoid refactoring. Similar reasons may apply to the case of code inspections. What is common to these two activities is that they do not produce new code, but rather lead to the improvement of existing code. Thus, they contribute long-term benefits rather than support the fulfillment of short-term needs. Programmers who do not appreciate this long-term value may express resistance when asked to perform these activities.

TASKS

1. How can pair programming (see Chapter 2) be conceived as code inspection?
2. Look at the Code Historian (*http://codehistorian.com/*) and Glance (*www.glance.net/site/Home.asp*) Web sites. What features for code inspections do these tools offer?
3. If you have to design a computational tool for code inspection, what features would you include in it? Think about people's needs and activities during a code review process.
4. Ask several software developers about the code inspections habits and procedures used in their companies. Summarize these habits and procedures. Can they be improved?

5. Since code inspection is done by the entire team, it may raise social issues. Suggest several of these issues and explain them.

REFACTORING

Study Questions

1. When are you upset by your code? What do you do in such cases?
2. Your team leader asks you to improve your code readability. How would you react?
3. In one of your team meetings, one of the teammates declares: "What I care about in software development is that the code runs. As soon as my code passes all tests, I leave it and do not improve either its structure or its readability." How would you react?

It is clear that the development of software is based on an iterative process through which the code structure is improved and other factors that influence its readability and comprehension are incorporated. Different software development methods deal differently with refactoring. For example, eXtreme Programming (XP) (see Chapter 2) includes refactoring as one of its 12 practices.

According to Fowler [Fowler00], refactoring is the process of changing a software code in such a way that although the observable behavior of the code is not altered, its internal structure is improved. When programmers refactor their code is not altered they usually remove duplications, improve communication, improve program comprehension, and add simplicity and flexibility. Accordingly, Kent Beck (in [Fowler00]) explains that the value that refactoring adds to a running program is expressed in the qualities that enable the team to continue developing at speed.

Refactoring aims to improve program comprehension mainly by improving the code design. For example, some refactoring actions lead to the elimination of duplicated code, so the code says everything once and only once. Although it may sound counterintuitive, refactoring helps us program faster because it leads to the production of code that is easier to work with and has fewer bugs and no patches.

An illustrative example for refactoring should be a relatively big computer program. However, as it is beyond the scope of this chapter, we do not base our discussion about refactoring on an example but rather on connections that refactoring has with human aspects of software engineering. Illustrative examples of code refactoring can be found in *Refactoring: Improving the Design of Existing Code* [Fowler00].

TASK

Look at the Refactoring home page (*www.refactoring.com*) and learn the main refactoring actions presented on the Web site. Refactor a program you worked on recently.

The reflective practice perspective (presented in Chapter 10) invites you to reflect on what you have done. Here are some reflection questions that can guide any reflective process that follows a refactoring session:

- What refactoring actions did you do and why?
- Is the code more comprehensible now? In what sense is the code improved now? What development actions can be performed easily now that could not be carried out before?
- Why was this refactoring conducted? In what ways might it help other developers to work with the code in the future?
- How did you know what to refactor? What bothered you in the code?
- Is there an essential difference between the two code designs (before and after the refactoring process)? If yes, what is it?
- Couldn't you write the refactored code when you wrote it in the first time?

Let us focus on the last question: "Couldn't you write the refactored code when you wrote it in the first time?" In many cases the answer to this question is "no." In this sense, the development of computer programs is similar to the way architects and other designers develop their creations. That is, only after some work is done is it possible to examine the work from the outside and to observe whether it can be improved. This is an important fact that software developers should be aware of, as it reduces their ambitions to create perfect code the first time they write it. In this spirit, Kent Beck says that refactoring reflects a new type of relationship with computer programs. Accordingly, when one really understands refactoring, the design of the system is as fluid and moldable in one's fingers as each of the individual characters in a source code file. This skill enables one to see how the code might change and be reshaped in a way that improves its structure (Kent Beck in [Fowler00]).

Some of the most relevant questions to be asked about refactoring are: If one wants to refactor the code, how would one find what to refactor? What clues in the code may help in such cases? Fowler [Fowler00] describes places in the code that require refactoring as places with a bad smell. For each of them, he suggests how to remove the bad smell. For example, in the case of duplicated code (when the same code structure appears in more than one place), Fowler suggests applying Extract Method; the same is recommended in the case of Long Method. However, in the case of Long

Method you are recommended to make a new method from the parts of the long method that seem to go well together.

Now that you are familiar with several refactoring activities, the question is, how do you manage the refactoring process? In other words, how is refactoring carried out? Based on his personal experience, Rasmusson [Rasmusson02] says that the team must refactor all the time, to the fullest extent. In the case described in that report, it is declared explicitly that when the team didn't follow this rule, the code became more cumbersome to work with. In practice, it means that in most cases, refactoring is conducted in small and local places, and sometimes a sequence of refactoring actions need to be performed.

TASK

Give an example of a sequence of refactoring actions. In what sense does this chain of refactoring actions improve the code?

Refactoring is not a simple task. Instead of continuing to work on your development tasks, you are asked to stop producing new code and continue working on a code that passes all tests, only to improve its quality. In many cases, you are not even sure that you will be one of the developers who will benefit from these improvements. Furthermore, refactoring requires a high level of awareness. Namely, if you are not aware of the fact that you should refactor, you will end up without refactoring. In addition, it is not sufficient to refactor a certain number of times. Rather, refactoring should be interwoven into the entire development process. Metaphorically, we may say that you need to switch between two hats: adding functionality to the code and refactoring.

As with other human beings' habits, one refactors or not depending on the culture in which one lives and one's attitude toward refactoring. For example, for an XP team, refactoring is part of the method; it is accepted naturally; it is part of the development routine; and it doesn't feel like an overhead activity. Furthermore, refactoring is tightly connected to the other XP practices, such as unit testing and continuous integration. At the same time, in other development environments, software engineers may:

- Claim that refactoring is an overhead activity and that they are paid to add functionality to the code, not to reshape it;
- Express resistance from different reasons (Opdyke in [Fowler00]). First, as mentioned previously, refactoring should be executed when the code runs and all the tests pass. Thus, it may seem that time is wasted. Furthermore, refactoring may break the passed tests and additional work will have to be invested to fix the code so that all the tests will run again. Second, if the benefits are long

term, why exert effort now? In the long term, the developer might not be part of the team to reap the benefits (p. 382).

As can be observed, there are many reasons not to refactor. If programmers believe in refactoring, however, they can also refactor when they are not part of an XP team. In such a case, one should treat refactoring as part of the profession of software engineering. Indeed, several Integrated Development Environments (IDEs) offer a Refactoring menu that includes actions such as Extract, Rename, and so forth. (See, for example, IntelliJ IDEA at *www.intellij.com/idea/* and Eclipse at *www.eclipse.org/platform*). This is a clear sign that refactoring has become part of the profession of software engineering.[3]

From all that has been said so far, it seems that refactoring is not a simple task from effective, cognitive, and technical perspectives. Furthermore, its benefits are not immediate and it requires a high level of awareness. However, the main message of this section is that programmers should not skip refactoring, partially because software development is a process that cannot be envisioned in advance and the need to reshape our products cannot be neglected. In this case, the analogy to architectural design is clear. Architects sketch many design drafts before they arrive at the final version of their creation. Many details and design decisions cannot be predicted in advance, as they become clear only as the design and coding process proceeds.

Another message of this section is that refactoring may improve the developers' programming skills. When programmers reshape the developed software through refactoring actions, they also improve their understanding of software development heuristics.

Discussion

- Discuss connections between the three main topics discussed in this chapter: program comprehension, code inspection, and refactoring.
- Connect the topics discussed in this chapter to the human nature of software teams (people come and go, developers want to gain "job security," among other factors).

SUMMARY QUESTIONS

1. Identify three difficulties with respect to each topic discussed in this chapter. Are all these difficulties of the same kind?
2. Identify three benefits of code inspections and refactoring. In what sense are these benefits similar? In what sense are they different?

FOR FURTHER REVIEW

1. Interview five software developers about their experience with program comprehension. Identify their difficulties. Ask them about their active attempts to improve the comprehension of the code they write. If they do not dedicate any attention to it, ask them why. Ask their opinion about what can be done to improve program comprehension.
2. Suppose you are asked to prepare a presentation about refactoring. You have access to all the available resources about refactoring. Prepare the skeleton of such a talk. Answer the following questions:
   - What is your main message?
   - What subtopics do you address? Examine these subtopics. What aspects of refactoring does each emphasize?

REFERENCES AND ADDITIONAL RESOURCES

[Biggerstaff94] Biggerstaff, Mitbander; Mitbander, Bharat; and Webster, Dallas, "Program Understanding and the Concept Assignment Problem," *Communication of the ACM* 37(5), (May 1994): pp. 72–82.

[Brooks83] Brooks, Ruven, "Towards a Theory of the Comprehension of Computer Programs," *International Journal of Man-Machine Studies* 18 (1983): pp. 543–554.

[Davies93] Davies, Simon P., "Models and Theories of Programming Theory," *International Journal of Man-Machine Studies* 39 (1993): pp. 237–267.

[Fowler00] Fowler, Martin (with contributions by Kent Beck, John Brant, William Opdyke, and Don Roberts) *Refactoring: Improving the Design of Existing Code*, Addison-Wesley, 2000.

[Fjeldstad and Hamlen83] Fjeldstad, R. K., and Hamlen, W. T., "Application Program Maintenance Study—Reports to Our Respondents," In *Tutorial of Software Maintenance*. (G. Parikh and N. Zvegintzov, eds.), IEEE Computer Society Press, 1983.

[Littman87] Littman, David. C.; Pinto, Jeannine; Letovsky, Stanley; and Soloway, Elliot, "Mental Models and Software Maintenance," *The Journal of Systems and Software* 7 (1987): pp. 341–355.

[Oman90] Oman, Paul W., "Topographic Style is More Than Cosmetic." *Communications of the ACM* 33 (5), (May 1990): pp. 506–520.

[Rasmusson02] Rasmusson, Jonathan, "Strategies for Introducing XP to New Client Sites," *Proceedings of the XP/Agile Universe Conference 2002* (2002): pp. 45–51.

[Vans99] Vans, A. Marie; von Mayrhauser, and Somlo, Gabriel, "Program Understanding Behavior During Corrective Maintenance of Large-scale Software," *International Journal of Human-Computer Studies* 51 (1999): pp. 31–70.

[vonMayrhauser93] von Mayrhauser, Anneliese, and Vans, A. M., "Identification of Dynamic Comprehension Processes During Large-scale Maintenance," *IEEE Transactions of Software Engineering* 22(6) (1993): pp. 424–437.

Deimel, Lionel E., and Naveda, J. Fernando, *Reading Computer Programs: Instructor's Guide and Exercises*, Software Engineering Institute, Carnegie-Mellon University, 1990.

Letovsky, Stanley, "Cognitive Processes in Program Comprehension," *The Journal of Systems and Software* 7, (1987): pp. 325–339.

Rajlich, Václav, "Program Comprehension as a Learning Process," *Proceeding of the First IEEE International Conference on Cognitive Informatics*, IEEE Computer Society Press, Los Angeles, CA, 2002.

Refactoring home page: *www.refactoring.com*

ENDNOTES

[1] The URL of the Annotated Bibliography about Code Reading and Program Comprehension is *www2.umassd.edu/SWPI/ProcessBibliography/bibcodereading.html*.

[2] Macadamian states that it has made "this checklist public, for the use of software development teams implementing code review as part of their process. Please feel free to use this checklist as it is, or to change it to customize a code review checklist that reflects your development style." For more information about Macadamian's code review processes and policies, read "Part of Your Complete Breakfast: Code Review Is a Source of Essential Vitamins and Minerals" at *www.macadamian.com/codereview.htm#CommentsandCodingConventions*.

[3] Although refactoring has so many advantages, there are cases where it should not be carried out. One of these cases is when the code is a mess and it would be better to start developing it from the beginning.

10 Learning Processes in Software Engineering

In This Chapter

- Introduction
- Objectives
- Study Questions
- Relevance for Software Engineering
- Software Engineering as a Reflective Practice
- Learning Organizations
- Conclusion
- Summary Questions

INTRODUCTION

Everyone who works in the software industry must be fully aware of the fact that software engineering is a dynamic area and that one cannot work in the field without being updated on a regular basis. This chapter answers this need for ongoing learning and addresses learning processes in software engineering by discussing two concepts. First, the concept of reflective practice is introduced and its relevance to software engineering is examined. Then, the focus is placed on learning organizations, which are working environments that foster and support learning processes both at the individual and collective levels.

OBJECTIVES

- Readers will appreciate the importance of learning processes in software engineering in general and of a reflective mode of thinking in particular.
- Readers will identify situations in which they might learn from applying a reflective mode of thinking and communal learning processes.
- Readers will become familiar with individual and collective learning processes.

STUDY QUESTIONS

1. Describe three situations in software engineering in which developers can learn from their previous experience and can improve their performance based on such learning processes.
2. Suggest three topics related to software engineering that require team learning. What characterizes these topics?
3. Suggest three mechanisms that may help software developers set "learning from a previous experience" to be a regular habit.
4. Suggest three situations in which software developers would tend not to share their knowledge with their colleagues. Suggest ways to encourage software developers to share their knowledge with their teammates in such cases.
5. Suggest three situations that do encourage software developers to share their knowledge with their teammates. What characterizes these situations?
6. Search the Web for the concept "learning organization."
   a. In a few sentences, describe the essence of this concept.
   b. Suggest at least three mechanisms/procedures that may help a software house become a learning organization.
7. Find on the Web three software houses that describe themselves as learning organizations. What characterizes these organizations?

RELEVANCE FOR SOFTWARE ENGINEERING

This chapter focuses on learning processes in software development. Specifically, it starts by examining the profession of software engineering as a reflective practice. According to this perspective, software developers are encouraged to continuously reflect on what they create as well as on the creation process itself, and to learn from such reflections. Then, software development is put in a wider context and is examined from the perspective of learning organizations. This perspective highlights the importance of communal knowledge and learning processes. More specifically, learning organizations aim to improve their performance by integrating learning processes and activities into their daily routines. All together, this chapter aims to illustrate how software developers may learn and improve their performance by continuously reflecting on their ways of thinking as well as by learning from their colleagues and working environment.

Because the discipline of software engineering is updated on a regular basis, one is unable to learn all that is included in the field. Rather, one has to find ways to foster one's professional skills by learning (in addition to the material itself) about learning processes and the use of such processes for the enhancement of personal and collective learning. In addition, the success of many software development projects relies on learning from previous experience. The perspective, by which one learns from the cumulative learning (of oneself and of the collective), gets its importance mainly from these facts that refer to software projects' success and failure.

From what we just said, the relevance of the two topics—reflective practice and learning organization—to software engineering is quite clear. The main question we address in this chapter is how to achieve and apply these practices. An important related practice that we hope to convey in this chapter is team learning. Senge [Senge94] says that the process of learning collectively is unfamiliar to most people because it has nothing to do with "school-learning" (pp. 355–356). We believe that experienced software developers feel that the concept of team-based learning process fits very well into software organizations.

SOFTWARE ENGINEERING AS A REFLECTIVE PRACTICE

This section examines ways a reflective mode of thinking may improve the performance of some of the basic activities that software engineers usually accomplish. Specifically, we describe the reflective practice perspective and suggest specific ways such an approach may be interwoven within software engineering activities.[1]

The reflective practice perspective, first introduced by Donald Schön [Schön83] [Schön87], guides professional practitioners (such as architects, managers, musicians, and others) toward the examination and rethinking of their professional creations during and after the accomplishment of the creation process. The working assumption is that the reflection process improves both proficiency and performance within such professions. Analysis of the field of software engineering and the type of work that software engineers usually accomplish [Winograd96] supports the adoption of the reflective practice perspective for software engineering.

The two main books that present the reflective practice perspective are *The Reflective Practitioner* [Schön83] and *Educating the Reflective Practitioner* [Schön87]. While the first book presents professions in which reflective thinking is (or should be) inherent, the second book focuses on how to *educate* students of such professions to become reflective practitioners. In these two books, Schön analyzes the added advantages one may obtain from continuously examining one's practice and one's thinking about one's practice.[2]

Laying out the topics that are possible subjects for reflection in software engineering, we may start with the actual creations (the software systems), going through a reflection on the way algorithms are developed and used in software systems, and moving on to skill-related topics such as topics related to human-computer interaction, software development methods, ways of thinking, and so forth. In fact, it seems that we might end up with a rich collection of objects that can be subjects of thought. It might be the result of the fact that software design embraces many aspects such as safety, human interface, ergonomics, graphics, algorithms, and program structure [Singer94].

The main tool, which we will use here in our discussion about being a reflective practitioner, is the ladder of reflection. Schön [Schön87] describes the ladder as a vertical dimension according to which higher levels of activity are "meta" to those below. When one moves "up," in this sense, one moves from an activity to reflection on that activity; when one moves "down," one moves from reflection to an action that enacts that reflection. The levels of action and reflection on action are described metaphorically as rungs of a ladder. When one climbs up the ladder, one makes what has happened at the rung below an object of reflection (p. 114).

Schön analyzes ladders of reflection in the context of student-tutor dialogues taking place in the architecture studio. Hazzan [Hazzan02] expands the use of ladders of reflection to student-coach dialogue in a software studio and to individual work. The idea in both cases, as well as in the cases discussed in the continuation of this section, is to illustrate how one can improve one's understanding when reflection is interwoven into the software development process.

In what follows the focus is on the construction of ladders of reflection in different situations. The described cases aim to illustrate how ladders of reflection may promote one's comprehension of the relevant development issue and may lead to insights that eventually may save time and money. We start by illustrating a ladder of reflection that takes place when two programmers, participating in a pair programming session, attempt to formulate, through a reflection process, what heuristic they should use (see Table 10.1). All tables that describe ladders of reflection illustrate the participants' thinking/discourse at each reflection rung.

TABLE 10.1 A Ladder of Reflection: The Case of Pair Programming

Ladder Rungs	Pair Dialogue
Designing (a process of reflection-in-action)	A: I'm going to use a stack here. Does this make sense?
Description of designing (takes the form of description with: appreciation, advice, criticism, etc.)	B: Good question. Let's explore the nature of the algorithm. Do you remember that in the last retrospective session we discussed a similar problem? What was it about?
Reflection on description of designing (reflection on the meaning the other has constructed for a description he or she has given)	A: You are right. I'm trying to recall. We started by comparing the nature of two projects and concluded that the project we discussed in that retrospective session is similar to what we developed last year. After that, we did a lot of reuse.
Reflection on reflection on description of designing (the parties to the dialogue reflect on the dialogue itself)	B: And, I even remember more clearly that following this retrospective session, we decided to change the format of our retrospective session. But more specifically, on the code level, we decided to change the design. Let's try to think in these directions: redesign and reuse. I guess they will save us a lot of time eventually.

Looking at the various rows in Table 10.1, we can see that the subjects of reflection on each rung are objects of different complexity: While detailed elements are the focus on the first rungs, algorithms, ways of thinking, and development approaches are at the center of the later rungs. It is suggested that such a reflective mode of thinking may improve the comprehension of one's own thinking. As software development is based on team interaction, it is reasonable to assume that the more one is aware of mental processes and ways of thinking (of oneself and of the others), the more teamwork is improved.

In what follows, we construct a ladder of reflection step by step. In each step, we try to explain why and how we move up the ladder. Needless to say, any statement may lead to the construction of many ladders of reflection. Still, we hope that this illustration clarifies how a reflective practice perspective may improve activities that software developers usually accomplish.

This illustrative ladder of reflection is constructed during a code review session (see Chapter 9, "Program Comprehension, Code Inspections, and Refactoring"). It ended when the team decided to change the way it runs its code review sessions. This example also illustrates how simple questions that invite reflective processes may lead to decisions about work habits.

Suppose that a software team, which is well trained in constructing ladders of reflection, competes on the development of a huge application for a leading insurance company. If it wins, their startup may become a leader in the market. The financial implications are clear. As usual, the team conducts a weekly code review session. The code review session starts with Developer A saying, "In order to implement this function, I used the quick sort algorithm." On the next rung of the ladder, Developer B refers to what Developer A has said and asks, "What were the reasons that guided you to select this algorithm?" Developer A answers, "I just picked the first sort algorithm that came to mind. I usually do not ask myself such questions."

The team leader is shocked, moves to the next rung of the ladder, and says: "I must admit that my working assumption was that before each of you implements any particular algorithm you make sure that it is the best algorithm for the particular task at hand. I think that we should make sure that all the algorithms you used in this application are the best ones for their particular purposes. To explain what I mean I would like to give you an example of a project that was developed three years ago in which. . . ." The team leader continues describing how in a previous project the entire development process was replaced when it was realized that a wrong process had been used. Then, the team continues with the code review session, analyzing, for each task, the fitness of the algorithms it uses.

After the code review session had ended, all the code was improved without giving up the high programming standards they set in their previous code reviews. As a result, the application's performance and design were improved significantly. The team won the competition and became the leader in the market. The team leader asked Developer B to summarize what happened by moving up to the fourth rung of the ladder. Table 10.2 summarizes the full ladder of reflection constructed during this event.

The next code review sessions were managed differently. More attention was placed on explaining what reasons led to specific decisions. This habit also forced each team member to become more reflective about development processes. The team members became more experienced with explaining why they used particular algorithms and consequently improved their problem-solving abilities.

TABLE 10.2 A Ladder of Reflection: A Code Review Session

Ladder Rungs	A Dialogue During a Code Review Session
Designing (a process of reflection-in-action)	Developer A: To implement this function, I used the quick sort algorithm.
Description of designing (takes the form of description with appreciation, advice, criticism, etc.)	Developer B: What were the reasons that guided you to select this algorithm? Developers A: I just picked the first sort algorithm that came to mind. I usually do not ask myself such questions.
Reflection on description of designing (reflection on the meaning the other has constructed for a description he or she has given)	Team leader: I must admit that my working assumption was that before each of you implements a particular algorithm, you make sure that it is the best algorithm for the particular task at hand. I think that we should make sure that all the algorithms you used in this application are the best ones for their particular purposes. To explain what I mean, I would like to give you an example of a project that was developed three years ago in which....
Reflection on reflection on description of designing (the parties to the dialogue reflect on the dialogue itself)	(after several days) Developer B: I was thinking of what has happened during the last couple of days and I think that we all should learn from it. One important implication would be that in our next code review sessions, each of us has to include a statement that explains why each algorithm is used and what alternatives were considered.

TASK

The ladder just described is only one possible ladder that can be constructed for this case. Suggest two other ladders that start with the first statement on the first rung; each ladder leads to different conclusions.

Ladders of reflection are not restricted to the use of software developers. This tool can also be used with other parties who are involved in the development process, such as the customer. The use of ladders of reflection in such cases is quite important; the literature is full of evidence of crises in software development processes that in many cases result from a misunderstanding between clients and software developers. A reflective mode of thinking, like the one suggested by the reflective practice perspective, may improve one's ability to understand the conceptions held by others in general and customers' needs in particular.

The following scenario takes place in a software development environment in which the customer is part of the development environment (such as eXtreme Programming (XP), see Chapter 2, "Software Engineering Methods"). Since the customer is on-site for answering questions, it is suggested that when both customer and developers are guided by a reflective mode of thinking, developers, as well as customers, may improve their understanding of the developed application. Table 10.3 presents a ladder of reflection that illustrates how this might happen when the customer, together with the development team, defines the next release/iteration. As described in Table 10.3, sometimes a reflective mode of thinking leads to wider implications than it aims at originally.

TABLE 10.3 A Ladder of Reflection: A Customer-Developers Dialogue

Ladder Rungs	A Conversation During a Planning Session
Designing (a process of reflection-in-action)	Customer: I want this feature to enable me to get a clear picture of my employees' opinions. You see, I consider my employees' opinions and experience when I think about the company's strategy. I truly believe that their viewpoint is very important.
	Developer 1: Can you think of a specific situation in which you will use this feature?
Description of designing (takes the form of description with appreciation, advice, criticism, etc.)	Customer: What do you mean? Would you like me to think about a specific case in which I will use it? A specific situation? Interesting. Yes, I can do that. If, for example, the marketing people tell me that one of our customers wants a specific feature, I ask the employees' opinion about the usefulness of the said feature. If I see that the majority tends to view it as a useful feature, sometimes I consider integrating it into our other products as well.

TABLE 10.3 A Ladder of Reflection: A Customer-Developers Dialogue *(continued)*

Ladder Rungs	A Conversation During a Planning Session
Reflection on description of designing (reflection on the meaning the other has constructed for a description he or she has given)	Developer 2: Can you specify the employees who usually respond to your questions? Customer: Sure, they all have a technical background, know the system very well, and are very loyal workers. I guess this is partially because of the good and supportive atmosphere that characterizes our company. Developer 3: I do not understand. Don't you ask any of the end users about their opinion with respect to the new feature? Customer: No . . . why should I? I fully trust my employees' opinion. But, now that you are asking, let me tell you how we were misled once in the past. Based on such a survey, we concluded that our customers need a very sophisticated feature. We spent about two development months on this development. During these two months, another company released a new product and (even worse) it turned out that this feature was useless. It is clear that we did not gain any competitive advantage from this bad decision. Fortunately, our customers did not leave us. The excellent service of our customer support is well known.
Reflection on reflection on description of designing (the parties to the dialogue reflect on the dialogue itself)	Customer: Wow! Let's rethink this feature that I started describing before and develop it in a way that our end users will be able to use it as well. That is, we will be able to also get the end-users' opinion. I'm so thankful. It is a real twist in my thinking about the introduction of new features to our system.

As can be observed, the customer understanding of what should be developed was improved throughout the session described in Table 10.3. In fact, the customer got the relevant insight only after a reflection mode of thinking was introduced. In this case, the reflection addressed similar situations in the past. We don't argue that such conclusions would not have been reached if customers hadn't been asked to reflect on past experiences. However, it is plausible to assume that such an insight would have arrived later (maybe only after a wrong feature was developed). The contribution of such lessons to software development processes is clear.

Conclusion

This section suggests adding the practice of reflection to software engineering activities. A similar approach is taken by Kerth's framework for project retrospective [Kerth01]. It is argued that a reflective mode of thinking may improve software developers' understanding of their and their teammates' ways of thinking. It is suggested that as a result, developers may improve their understanding of their own way of software development as well as their understanding of the development environment.

It is important to note that since reflection does not produce new code, some developers (and team leaders) may not see the benefits of using it. Thus, to be carried out successfully, reflection should be supported by the development environment. A learning organization (discussed in the continuation of this chapter) is one environment that may support, and even promote, a reflective mode of thinking. Learning organizations do not limit their activities to those that yield short-term profits. Rather, energy and resources, which support activities that lead to long-term targets, are invested and allocated as well.

Activities

Select a topic to be the subject for the ladder of reflection. Separate into pairs. Work on the following tasks:

1. Each pair constructs a ladder of reflection with respect to the selected topic. After all pairs complete the construction of their ladders, compare the various ladders.
2. All pairs are given an opening sentence to be presented on the first rung of the reflection ladder. Each pair composes the upper rungs of the ladder. After all pairs construct their ladder, compare the ladders. Discuss similarities and differences among the ladders.

Discussion

1. How can a software team establish a development environment in which a reflective mode of thinking is an inherent part?
2. How should team members respond to one team member who refuses to participate in reflection processes but who has a central role in the development software?

LEARNING ORGANIZATIONS

This section deals with the concept of learning organizations in general and its application in the context of software organizations in particular. The idea of a learning organization acknowledges the fact that knowledge and information are essential assets in software organizations and should be managed as tangible assets. Many reasons support the creation of a learning organization (gaining competitive advantage, for example). One of their common aims is to tie the individual professional development of every employee to the superior economic performance of the organization ([Senge94], p. 10). This linkage is achieved by the continuous analysis of past experience and the translation of that accumulative experience into knowledge that is accessible by the entire organization, and relevant to its core purpose ([Senge94], p. 49). This basic characteristic of learning organizations is logical when we think about software learning organizations, where learning from previous experience may have direct influence on the success of future projects.

The term *learning organization* was coined by Peter Senge in his book *The Fifth Discipline* [Senge90]. According to Senge [Senge90], learning organizations are organizations where people continually expand their capacity to create the results they truly desire and where people are continually learning to see the whole together (p. 3). In practice, learning organizations provide their employees with a working environment in which learning is an integral part of everyday work and routines. The atmosphere in such organization rewards (not necessarily financially) employees who improve their performance, promote teamwork, and enhance the organizational targets.

The term *learning organization* does not refer to a specific type of organization (hospitals, software houses, shoe manufactories, or design offices, for example) or to a specific organizational structure. Rather, a learning organization is identified by the fact that individual and collective learning are intertwined in order to improve the organization's performance. It is important to note that being a learning organization is part of the organizational culture, and thus management support in such learning-oriented activities is essential.

TASK

Skyrme [Skyrme99] says that in addition to the collective and the individual learning that characterizes learning organization, learning organizations are adaptive to their external environment and continually enhance their capability to change and to adapt to that environment. Explain how this characteristic of learning organizations may be reflected in the daily life of software developers.

A learning organization gives its employees the feeling that their *personal* learning influences the organization development and that the synergy principle is applied also with respect to individual knowledge. In other words, in learning organizations it is believed that the contribution of the accumulative learning to the achievement of organizational targets is bigger than the potential contribution of the sum of the individual's knowledge. With this respect, Hedberg [Hedberg81] says that although organizational learning occurs through individuals, organizational learning is more than the commutative result of its members' learning. Metaphorically, Hedberg says that organizations do not have brains, but they have cognitive systems and memories. Accordingly, just as individuals develop personalities, organizations develop world views and ideologies. Thus, although members of the organization may come and go, organizations' memories preserve behaviors, norms, and values over time (page 6, in [Fiol and Lyles85], p. 804).

TASK

Suggest strategies that organizations may use to enhance organizational learning. (This issue is addressed later in this chapter.)

According to Senge's theory, a learning organization is based on five principles, each of which creates one dimension of learning organizations. They are described in brief here:

> **Systems Thinking:** Systems thinking is a conceptual framework based on patterns (may be complicated) that may explain different events as instances of one phenomenon. Systems thinking is about abstraction (see Chapter 11, "Abstraction and Other Heuristics of Software Development"), since it helps us ignore details and examine what different events have in common. Such a perspective may improve our analysis abilities and may guide us in decision-making processes. Needless to say, such understanding is based on learning, examination, and reflection of what has happened in the past.

TASK

Give three examples related to software engineering that may fall under the umbrella of systems thinking.

> **Personal Mastery:** Personal mastery means learning to expand our personal capability to create the results we desire, together with creating an organizational environment in which members are encouraged to develop themselves toward the goals and purposes they choose ([Senge94], p. 6). Clearly, personal mastery connects the individual's learning with the learning of the organization.

TASK

Discuss connections between the personal mastery dimension and the Code of Ethics of Software Engineering, discussed in Chapter 5, "Code of Ethics of Software Engineering."

> **Mental Models:** Mental models is a concept used in other learning theories as well. The discipline of mental models guides us to improve our understanding of our own mental models that guide our behavior and our conception of the world. By doing so, we may examine the nature, advantages, and disadvantages of the process we go through. Based on the lessons learned we may conclude about what future directions to adopt and follow.

TASKS

1. Reflect on one of your previous experiences in software development in which you realized that your mental model of some element in your development environment seemed to be wrong.
2. Discuss how the Reflective Practice perspective, described early in this chapter, may support the discipline of Mental Model.

> **Shared Vision:** Shared vision means building commitment in a group, by the development of shared images of the future the group seeks to create, and the principles and guiding practices by which the group hopes to get there ([Senge90], p. 6).

TASKS

1. Suggest two cases in software development. In one case, a shared vision leads to a big success; in the second case, the lack of a shared vision leads to the company failure.
2. Is it possible that a company with a shared vision fails? If yes, how can this phenomenon be explained within the learning organization framework?

Team Learning: The discipline of team learning combines many of the elements described in Chapter 3, "Working in Teams": listening, knowledge sharing, patterns of interaction, and the way the team operates. Ideologically, Team Learning means the transformation of collective thinking skills in a way that enables groups of people to develop an ability that is greater than the sum of individual member's talents ([Senge94], p. 6).

TASKS

1. Discuss connections between this dimension of learning organizations, software engineering methods (Chapter 2), and teamwork in software development (Chapter 3).
2. Suggest two activities that can foster learning in teams. What characterizes these activities? In what ways do these activities enhance team learning?
3. For each of the following activities, explain its potential contribution to the establishment of a learning organization:
   a. Encourage team members to raise any problem they face.
   b. Inspire the idea that not all problems have one unique solution.
   c. Invite suggestions for improvements and new development directions.
   d. Set brainstorming (thinking "outside of the box") sessions.

The preceding questions lead naturally to the discussion about the ability to share and leverage knowledge across the organization. A closely related topic is knowledge management. Knowledge management is vital in learning organizations, but it is difficult to manage knowledge. Although everyone agrees that knowledge management is about knowledge, there is no agreement on one unique definition.

The following brief explanation of the concept of knowledge management is based on the definition presented on the Knowledge Management Research Center Web site (*www.cio.com/research/knowledge/edit/kmabcs.html*). There is no agreement upon the definition of for knowledge management, but most definitions concur that knowledge management is the process and activities through which organizations generate value from their intellectual and knowledge-based assets. Among other things, knowledge management is based in activities that support knowledge sharing among employees, departments, and even other companies, all in an effort to create best practices. It's important to note that the definition says nothing about technology; while knowledge management is often facilitated by IT, technology by itself is not knowledge management.

TASK

Based on a Web search, find:

- Examples of learning organizations.
- Examples of knowledge management tools.
- A successful story about knowledge management.
- A story about a failure in knowledge management.
- Mechanisms that encourage learning cycles and getting feedback.
- Ways to establish learning organizations. Are all these ways technology based?

Not surprisingly, technology may play a significant role in learning organizations. Among other tools, the Internet and intranets may support knowledge management processes as they invite access to information and knowledge sharing. That means, for example, that if an individual wants to access a specific piece of information (and is authorized to access it), he can access it without asking the permission of anyone who may have some difficulties with information sharing. It is easy to observe how such an environment fosters mutual information exchange and creates a suitable infrastructure for a learning organization.

Another useful tool for a learning organization is electronic discussion groups. In learning organizations, it is natural to create electronic discussion groups to foster focused discussions. This is especially relevant when a new initiative is developed and a rapid exchange of information is needed. The advantage of this type of information exchange over face-to-face interaction is that electronic discussion groups enable asynchronic interaction in time and place. It is also a useful tool when people need some time to grasp new ideas. Needless to say, face-to-face communication should not be given up; rather, the two types of interaction should be integrated.

TASK

Suggest three topics related to software development that are conducive to being discussed electronically.

The Internet, intranets, and discussion groups are only examples of technological tools that may support the creation of learning organizations. It is important to note that it is not argued that the moment the technological infrastructure exists in an organization, the organization becomes a learning organization. What is argued is that technology is a means that can support and enhance the creation of a learning organization culture.

So far, we have discussed the concept of learning organization in general. With respect to the profession of software development, the paradigm of learning organization is reflected in the term *learning software organizations.* The importance of being a learning *software* organization is derived from the fact that software development is a competitive industry and new players can enter it with only the asset of *knowledge* (without any physical asset). This, together with the acknowledgment that knowledge and learning are main players in this industry, adds to the importance of the concept of learning organization in the software industry.

TASKS

1. Suggest a learning mechanism whose aim is to improve the quality of software.
2. Suggest a learning mechanism whose aim is to shorten software time to market.

Activity

Suppose you establish a software startup. Work on the following tasks:

1. Describe the startup.
2. Lay out the basic activities you would set up to make it a learning organization.
3. Discuss what might happen if these activities are not set up when the startup is established, but rather a year later.

Discussion

Integrate the two subsections in this chapter (software engineering as a reflective practice and learning organizations) and suggest specific mechanisms by which reflective processes can be interwoven into learning organizations.

CONCLUSIONS

This chapter starts with the examination of software engineering as a reflective practice. Specifically, it suggests the construction of ladders of reflection as a means to help software developers improve their understanding of the development environment. It is suggested that such an experience may be naturally integrated into the framework of learning organizations. In practice, becoming a learning organization requires that each individual adopt a perspective and work habits that are open to new opinions and changes. In the software industry, these characteristics are vital for a company to survive.

SUMMARY QUESTIONS

1. What benefits might one gain from reflective processes?
2. What situations are appropriate for reflective processes?
3. In what cases might reflection be waste of time?
4. Suggest connections between the following concepts: knowledge management, information technologies, and learning organizations.

FOR FURTHER REVIEW

1. For each of the two cases described below, address the following:
    a. Identify potential topics for reflection.
    b. Construct a ladder of reflection.
    c. Describe what lessons you learned from this experience.
   Case 1—Testing in eXtreme Programming: In XP [Beck00], programmers write tests before they write the code. This approach is called test-driven-development (TDD). Write a class `Student` according to the TDD approach.
   Case 2—New developers join your team: You are a team leader. You are asked to accept to your team two software engineers who have developed software that can be integrated into the project on which your team works. You ask your team and the two new developers to meet to discuss how to merge the code successfully.
2. In this question, you are asked to apply the ladder of reflection process to a case in which it is carried out by an individual. The idea is to illustrate how one can increase the level of abstraction of one's thinking when

reflection is interwoven in the process of software development. In Table 10.4 all phrases are a developer's thoughts.

TABLE 10.4 A Ladder of Reflection: Individual Work

Ladder Rungs	Developer's Thoughts
Designing	I do not know if this should be a class or an attribute.
Description of designing	This is a crucial decision for the continuation. Before I move on, it would be better if I check the implications of each decision.
Reflection on description of designing	Okay, now that I know the implications of each decision, how, in practice, can I determine between the two options? What guidelines may help in such dilemmas?
Reflection on reflection on description of designing	I got it. It should be an attribute. It is interesting, because at first glance it seemed to be a class. I must understand how the guidelines I used helped me to arrive to this conclusion.

Assume a situation in which a programmer looks for a way to decide whether all the needed tests were conducted. Construct two ladders of reflection for such a situation.

3. Senge's field book [Senge94] presents vast collection of tools, procedures, and heuristics to foster team learning. Select two of these methods and apply them to situations in which you should solve real problems. During and after that process, reflect on the following issues: In what ways did the use of these methods help you in solving the problem? In what ways do these methods differ from other methods for team discussion you used to use in problem-solving situations? Can you suggest improvements for these methods?

4. In the context of the industry/academia communication chasm, Glass [Glass97] says that industrial people tend to reinvent the same wheel they invented last year (p. 13). Suggest real-life situations in which practitioners'

reflection on the way they solve problems may help them learn from previous failures as well as from previous successes.
5. Tape record one of your team meetings whose aim is to solve a problem. Listen to the cassette and analyze the meeting: Did all participants contribute to the discussion? Did someone discourage the introduction of new ideas? In what places would you navigate the meeting differently? Illustrate your analysis by quoting excerpts from the meeting. Summarize: Did the meeting achieve its aims? Could it be managed more efficiently?

REFERENCES AND ADDITIONAL RESOURCES

[Beck00] Beck, Kent, *Extreme Programming Explained: Embrace Change*, Addison-Wesley, 2000.

[Fiol and Lyles85] Fiol, C. Marlene, and Lyles, Marjorie A., "Organizational Learning," *Academy of Management Review* 10(4), (October 1985): pp. 803–813.

[Fowler02] Fowler, Martin, "The New Methodology," *martinfowler.com*, www.martinfowler.com/articles/newMethodology.html, 2002.

[Gates99] Gates, Bill, with Hemingway, Collins, *Business @ the Speed of Thought: Using a Digital Nervous System*, Penguin, 1999.

[Glass97] Glass, Robert L., "Revisiting the Industry/Academe Communication Chasm," *Communication of the ACM* 40 (6), (June 1997): pp. 11–13.

[Hazzan02] Hazzan, Orit, "The Reflective Practitioner Perspective in Software Engineering Education," *The Journal of Systems and Software* 63(3), (2002): pp. 161–171.

[Hazzan and Tomayko03] Hazzan, Orit, and Tomayko, Jim, "The Reflective Practitioner Perspective in eXtreme Programming," *Proceedings of the XP Agile Universe 2003*, New Orleans, Louisiana (August 2003): pp. 51–61.

[Hedberg81] Hedberg, Bo, "How Organizations Learn and Unlearn." In P. C. Nystrom and W. H. Starbuck (eds.). *Handbook of Organizational Design*, Oxford University Press 1981: pp. 8–27.

[Kerth01] Kerth, Norman L., *Project Retrospective*, Dorest House Publication, 2001.

[Schön83] Schön, Donald A., *The Reflective Practitioner*, BasicBooks, 1983.

[Schön87] Schön, Donald A., *Educating the Reflective Practitioner: Towards a New Design for Teaching and Learning in the Profession*, San Francisco: Jossey-Bass, 1987.

[Schön96] Schön, Donald A., "Interviewed by John Bennent: Reflective conversation with Materials." Terry Winograd, *Bringing Design to Software*, Addison-Wesley, 1996: pp. 171–184.

[Senge90] Senge, Peter M., *The Fifth Discipline: The Art and Practice of the Learning Organization*, Currency Doubleday, 1990.

[Senge94] Senge, Peter M., *The Fifth Discipline: Fieldbook*, A Currency Book, published by Doubleday, 1994.

[Singer94] Singer, A., "Towards a Definition of Software Design," *Design+Software—The ASD Newsletter* (1994), *www.pcd.stanford.edu/asd/info/articles/definition.html*.

[Skyrme99] Skyrme, David, "The Learning Organization," *www.skyrme.com/insights/3lrnorg.htm*, 1999.

[Winograd96] Winograd, Terry (ed.), *Bringing Design to Software*, Addison-Wesley, 1996.

Peter Senge and the learning organization: *http://www.infed.org/thinkers/senge.htm*

Group Performance Systems, Inc., Learning Organization Resources on the Web and on the Net: *http://www.gpsi.com/lo.html*

Review of *The Fifth Discipline*: *http://www.rtis.com/nat/user/jfullerton/review/learning.htm*

The Knowledge Management Resource Center: *http://www.kmresource.com/*

ENDNOTES

[1] This section is partially based on [Hazzan02] and on [Hazzan and Tomayko03].

[2] Schön did not discuss in his two original books the application of the reflective practice perspective with respect to software engineering. He did address this issue in a later interview [Schön96].

11 Abstraction and Other Heuristics of Software Development

In This Chapter

- Introduction
- Objectives
- Study Questions
- Relevance for Software Engineering
- Central Heuristics of Software Development
- Additional Topics Related to Abstraction
- Summary Questions

INTRODUCTION

The process of software development comprises different activities, each with its own objective. Some activities aim to improve the code (e.g., testing and code review); others aim at improving the software from the customer's perspective (requirements definition).

This chapter examines different types of activities that are carried out during the software development processes. These activities are continuously conducted, yet it is not obvious specifically when they are performed. They are, in fact, heuristics (or ways of thinking) that one may employ when one performs other activities. One of these ideas is abstraction, discussed previously (see Chapter 8, "The History of Software Engineering," and Chapter 10, "Learning Processes in Software Engineering").

This chapter starts by describing three heuristics: structured programming, successive refinement, and abstraction. Then, the focus is placed on abstraction, and different chapters of the book are reexamined through the lens of abstraction. This examination is conducted to illustrate how abstraction may highlight additional dimensions of almost any topic connected to software engineering, and consequently how such an examination may deepen our understanding of this topic. We convey that abstraction is a perspective that can be used for different purposes in different situations. This message is based on the assumption that the awareness of the concept of abstraction may improve one's performance within the profession of software engineering.

The next concepts discussed in this chapter are the human aspects of software architecture and the use of metaphors in software engineering processes. This chapter concludes by describing how abstraction is expressed in computer science and software engineering academic programs.

OBJECTIVES

- Readers will become aware of heuristics that can guide the performance of different activities throughout the process of software development.
- Readers will become familiar with the idea of abstraction and its relevance and contribution to software development processes.
- Readers will increase their awareness of situations in software development in which thinking in terms of different levels of abstraction may improve and enhance software development processes.
- Readers will examine topics discussed elsewhere in the book from the perspective of abstraction.
- Readers will know what "software architecture" means and will be able to derive quality attributes from the software architecture to meet the desires of the user.
- Readers will know how a "metaphor" is used in software engineering.

STUDY QUESTIONS

1. Look at software engineering books for the concept of abstraction. Identify the main features of the concept as they are described in these books. Explain how these features contribute to software development processes.

2. Look at software engineering books for the concept of successive refinement. In what ways does successive refinement contribute to software development processes? What connections does it have to abstraction?
3. Look at Chapter 10 and examine the context abstraction is discussed in there. Suggest additional situations in software development that can benefit from developers' thinking in terms of different levels of abstraction.
4. What is the difference between architecture and design?

RELEVANCE FOR SOFTWARE ENGINEERING

This chapter discusses connections between principles (or heuristics) of software development and the human aspects of software engineering. Naturally, since heuristics are addressed, the focus is placed on the cognitive aspect of software engineering. Our main target in this chapter is to highlight ideas that can contribute to software development processes, independently of any particular software development method or programming language.

At the beginning of the chapter, we introduce briefly three heuristics: structured programming, successive refinement, and abstraction. Then, the focus is placed on abstraction—one of the central ideas of software development. As mentioned earlier, the relevance of abstraction to software engineering is illustrated by discussing different topics that have been discussed so far in the book from the perspective of abstraction.

The relevance of the software architecture to software engineering processes seems clear. With respect to software architecture, we discuss the concept of metaphor. In general, metaphors become a useful communication tool between customers and developers and among developers. Specifically, eXtreme Programming (XP) (see Chapter 2, "Software Engineering Methods") includes the use of metaphor as one of its core practices.

CENTRAL HEURISTICS OF SOFTWARE DEVELOPMENT

The cognitive complexity of software development is widely acknowledged, and the discipline of computer science (and later also the discipline of software engineering) developed heuristics to overcome this complexity. These heuristics became part of the discipline and are discussed explicitly in computer science and software engineering literature. In this chapter, we mention structured programming, successive refinement, and abstraction.

Structured Programming

The idea of structured programming was introduced by Dijkstra [Dijkstra72]. It was one of the first program design methods targeted at coping with the increasing cognitive complexity of developing software systems. It guides developers to break down big computer programs into routines that have clear roles in the whole computer program. Structured programming is accepted today as one of the fundamental principles of software development.

Structured programming is reflected in related heuristics employed in software development, such as top-down design and bottom-up programming. All these methods aim to keep the program within one's intellectual grasp. In practice, they all direct one to deal with ideas instead of details. Top-down design, for example, guides developers to start by breaking the problem into a set of subproblems and then to divide each subproblem into subproblems. The process continues in this manner until each subproblem is defined at a basic enough level and further decomposition is unnecessary.

TASK

Look at software engineering books for the principle of bottom-up design/programming. Explain its connections to top-down design/programming and in what ways it supports software development processes.

Successive Refinement

The idea of successive refinement is similar to the principle that guides the process of book writing. For example, in the case of this book, we started with a list of topics related to software engineering, that we found relevant from the human perspective. Here are three examples: teamwork, the Code of Ethics of Software Engineering, and learning processes in software engineering. Then we started elaborating each of these topics. This elaboration is was expressed first by the headings we added to each chapter. The content was added only in the next stages.

During that process, we sometimes changed the location of different topics and organized sections differently. However, at each moment, the full picture remained complete and only the details were added, changed, or removed. Adding details is carried out in different ways. Sometimes, all we have to do is to add the content to a section that was determined in early stages; sometimes a new section is added to complete the picture; in yet other cases, sections are broken into subsections. However, according to the idea of successive refinement, no matter what kind of elaboration is conducted throughout this process, the entire picture of the book, as we conceive of it at each stage, is completed.

A similar process occurs during software development. One starts by describing a solution to the problem. This description uses terms that capture the essence of the solution, but these terms are not necessarily provided by the programming language the software is developed in. Then, details are added to the solution in steps similar in nature to the stages mentioned in the previous paragraph. For example, sometimes only the description of a term is added; sometimes a term is broken down to several subterms. Dijkstra [Dijkstra72] describes the main idea of this process by saying that independent of the final decisions that are taken, the coding of the earlier levels remains valid. He concludes that in view of the requirement of program manageability, this continuity is veryencouraging (p. 39). Note that Dijkstra addressed explicitly the importance of program manageability, which aims to support the cognitive complexity of software engineering.

It is important to be aware of the idea of successive refinement when learning programming or when reading computer programs that someone else has written (as happens in many cases in software development). In both situations, computer programs are sometimes presented as finished products. Their aims and syntactic details are explained, but the process by which they were developed is not always discussed [Wirth71]. However, this style of presenting computer programs shows what computers can do, but does not reflect the actual process of software development. As a result, novices may get the incorrect impression that programming is based on syntactic skills only, and that writing a computer program is a simple process in which the developer's intuition is simply translated into the programming language.

One way to help novice software engineers grasp the actual process of software development is to explain the process through which the software was developed (see the process-oriented perspective of software development presented in Chapter 7, "Different Perspectives of Software Engineering"). It is more and more accepted that this type of documentation is required for program comprehension. Furthermore, it is more and more accepted that the specific ways by which the computer program achieves its goals should be transparent merely from the code itself.

In educational environments, Pattis [Pattis81] suggests presenting the students with the process through which computer programs are developed. This presentation, Pattis explains, may help students discover that one should not get carried away by the "I've got to get it perfect the first time" syndrome (p. 83). This syndrome, Pattis says, simply leads programmers to a situation in which they are unable to develop computer programs.

TASKS

1. Imagine you are asked to teach the idea of successive refinement. How would you teach it? What guidelines will you follow?
2. Discuss connections between successive refinement and structured programming.

Abstraction

Simply put, abstraction is a cognitive means according to which we concentrate on the essential features of our topic of thought, and ignore details that are not relevant at a specific stage of problem-solving situations. Abstraction is especially essential in solving complex problems, as it enables the problem solver to think in terms of conceptual ideas, rather than in terms of their details.

The concept of abstraction is an integral part of the professions of computer science and software engineering. Abstraction can be expressed in different ways, all guiding us to ignore irrelevant details at specific stages in order to overcome complexity. The next three paragraphs describe three ways by which abstraction can be expressed.

One way is by observing what a group of objects has in common and capturing this essence by one abstract concept. In such cases, abstraction leads us to preserve what is common to a set of objects and to ignore irrelevant differences among them. The common characteristics are captured in one concept (which can be a mathematical concept, a class in the paradigm of object-oriented development, a procedure that deals with different kinds of inputs, etc.). In this sense, abstraction is mapping from many to one. For example, the hierarchy of mammals reflects abstraction; at its higher levels we ignore differences between the mammals, at its lower levels, we distinguish between different characteristics of mammals. Another example is mathematical definitions. Think about a square, for example. The definition of square ignores the specific side's size and location. However, in lower levels of abstraction, when we discuss a specific square, we do care about these details. In general, any mathematical definition describes a concept by capturing their shared properties and ignoring irrelevant differences between them.

Abstraction is also applied by choosing appropriate language for describing a solution for a given problem. This language is not necessarily based on the tools provided by the programming language that we use. However, abstraction in this case helps developers think about the problem with appropriate conceptual terms, without being guided to think in terms of any particular programming language. In this case, without using abstraction, computer languages would have forced software developers to delve into irrelevant details at too early stages, and would, in fact, take control of the programming process. In this sense, abstraction bridges natural language and programming language. This idea is reflected by the evolution of programming languages. The first programming languages were significantly different from our language, but today's programming languages enable software developers to express ideas in a way that becomes more and more similar to natural language.

The third expression of abstraction is applied by describing objects by their characteristics rather than by the way they are constructed or how they work. Accordingly, Hoare [Hoare86] explains that an abstract command is a command that specifies the properties of the computer's desired behavior without prescribing in detail how that behavior will be achieved. This idea is expressed, for example, by writing one set of instructions for the manipulation of different types of objects. In lower levels of abstraction, the different ways in which these instructions work on different types of objects are specified. This idea is illustrated by Abelson and Sussman [Abelson and Sussman86] by the process of setting abstraction barriers (p. 73).

TASKS

1. Suggest specific situations in software development in which each of the aforementioned implementations of abstraction is useful. Explain what benefits one gains from each expression of abstraction.
2. Can these uses of abstraction be applied to other situations in our lives?
3. How is abstraction applied in object-oriented design?

ILLUMINATION OF PREVIOUS CHAPTERS BY ABSTRACTION

This section highlights several of the previous chapters of this book from the perspective of abstraction. By doing so, this section is in fact an application of one of the ideas presented in Chapter 7. In Chapter 7, we said that the richness of software engineering enables us to examine it from different perspectives. Here, the lens of abstraction is used.

The first chapter of this book deals with the nature of software engineering, emphasizing its human aspect. Cognitive and social aspects involved in software development are illustrated by describing two working days in the life of software developers. These stories illustrate "how much" of such days are about human aspects and not about technical aspects. In this sense, these narratives establish the rationale for the book. We use low levels of abstraction (details) to bring to the surface abstract topics, such as communication in teamwork. Such use of abstraction is appropriate when new concepts are introduced. In our case, the new idea is the human aspect of software engineering. In later chapters, some of the details described in Chapter 1, "The Nature of Software Engineering," are addressed at higher levels of abstraction.

TASK

Specify at least five details related to the human aspect of software engineering that are mentioned in the stories presented in Chapter 1 and are discussed in later chapters at higher levels of abstraction.

Chapter 2 addresses software development methods from the human perspective. It aims to illustrate that both technical and human factors should be considered when one evaluates what software development method to adopt for a specific software project. Specifically, three methods are examined: the Spiral Model, the Unified Process, and eXtreme Programming. For each of these methods we describe how the four basic activities of the paradigm of software engineering—specifying, designing, coding, and testing—are implemented. The rationale for this set of activities, on which most of the accepted software development methods are based, is explained in Chapter 8.

Abstraction is expressed, in this case, in the term *paradigm*. Paradigm reflects the fact that we capture the shared essence of the activities of which the paradigm consists, while ignoring the details of how each of these activities is implemented by each software development method.

TASKS

1. With respect to each of the three software development methods described in Chapter 2, explain the specific implementations of the paradigm of software engineering.
2. The following statements were said by imaginary software engineers. In your opinion, what is common to all of them?
   - "I need to gain a global view of the application in order to know how this method fits into it."
   - "I truly believe that if I had a minute to think about these two objects more abstractly, I'd come up with the conclusion that they can be extracted into one class. But I must move on to the next development task."

- "I need some time to think about the code without being swamped with all the details. I'm almost sure that if I could leave now and go jogging, I'd come up with a solution. But I must stay as late as all the others on my team."
- "I wish I could join the programmers when they write the code. You ask why? I'm not sure if this complicated design can be implemented into C++."

3. During the process of software development, developers should think in terms of different levels of abstraction and move between abstraction levels. For example, when trying to understand customers' requirements during the first stage of development, developers must have a global view of the application (high level of abstraction). Conversely, when a developer codes a specific class, a local perspective (on a lower abstraction level) should be adopted. Obviously, there are additional levels of abstraction in between these two levels of abstraction.
   - Describe different situations (or activities) in software engineering that require one to think in terms of different levels of abstraction.
   - Review the different XP practices and identify those practices that guide software developers to think in terms of different levels of abstraction when appropriate.

Chapter 3, "Working in Teams," discusses different topics related to software teamwork, one of which is the structure of software teams. When we discuss hierarchical teams, we refer to the influence of hierarchies on teamwork in the context of software development. Among other influences of hierarchies, we say that "Grunts" in such a team organization may miss the big picture of what goes on. Consequently, they may have only a narrow understanding of the development environment in general and the developed software in particular. From the perspective of abstraction, these programmers are familiar only with the details and cannot conceive of the developed application from a higher level of abstraction. This narrow perspective limits their understanding of the development activities. For example, they may not understand how the details are connected to each other and how the big picture is composed of them.

An appropriate metaphor for this situation is one's familiarity level with a new city. If one knows only the details of how to arrive from one specific location to another specific location, one's ability to cope with new situations remains limited. Such local understanding may not be sufficient, for example, for the construction of a new path for which one does not have specific instructions. When one has a more global and abstract (less detailed) image of the city, the ability to cope with new situations is improved significantly. In such cases, when one is familiar with connections between different parts of the city, but does not know all the details of all its parts, one may perform better when the need to construct new paths emerges.

We suggest that developers' familiarity with the big picture of the developed application may improve their performance in developing specific tasks. One way to achieve this familiarity with the entire picture of the developed application is by the Planning Game (one of the XP practices; see Chapter 2 and Chapter 13, "Software Project Estimation and Tracking"). Although each developer is responsible for specific tasks, they all participate in the Planning Game. Playing the Planning Game enables them to become familiar with the entire picture of the developed application. In later development stages, when they have to make decisions related to other parts of the application, this knowledge may be useful.

Chapter 3 also discusses dilemmas that may arise during software development processes. For example, the Prisoner's Dilemma is presented to explain why people tend to compete even in situations when they might gain more from cooperation. Abstraction is expressed in this case since many real-life situations are captured in one theoretical framework—the Prisoner's Dilemma. In practice, when we face a problem of a similar nature, this abstraction may guide us to ignore the details of the specific situation we are faced with and to consider the solution that this framework offers.

TASK

1. Describe three situations related to software engineering that can be described within the framework of the Prisoner's Dilemma. What features do they have in common? In what sense are they different?
2. This task continues the city metaphor just presented. Describe how abstraction is expressed in geographical maps. In what ways is this application of abstraction similar to the application of abstraction in software development processes? In what sense are these two applications of abstraction different?
3. Suggest an example that illustrates how developer's familiarity with the entire picture of the developed application influences the development of one specific task for which the programmer is responsible.

Abstraction is expressed also in Chapter 4, "Software as a Product," in Chapter 6, "International Perspective on Software Engineering," and in Chapters 7 and 8. With respect to customers (Chapter 4), the customer's requirements themselves are an expression of abstraction. This is because we do not discuss the details of how these requirements are implemented but rather we stay in higher levels of abstraction and describe only the essence of the requirements. With respect to the international perspective on software engineering (Chapter 6), abstraction is expressed by addressing cultural issues. The concept of culture itself is an abstraction, as it captures the similar behavior of many people, ignoring differences among individuals. Chapter 7, which describes different perspectives on software engineering, illustrates abstraction by analyzing properties of the profession of software engineering rather than describing details and procedure. Chapter 8 describes how abstraction became part of the paradigm of software engineering.

Chapter 5 deals with the Code of Ethics of Software Engineering. Not only is the concept of ethics abstract, abstraction is expressed by how the Code of Ethics of Software Engineering is presented. The preamble of the Code of Ethics of Software Engineering, states explicitly that "[t]he short version of the code summarizes aspirations at a high level of abstraction. The clauses that are included in the full version give examples and details of how these aspirations change the way we act as software engineering professionals. Without the aspirations, the details can become legalistic and tedious; without the details, the aspirations can become high sounding but empty; together, the aspirations and the details form a cohesive code."[1]

The essence of the Code of Ethics of Software Engineering is discussed in Chapter 5 through the analysis of different scenarios taken from the world of software development. This analysis can be characterized as an examination of the code from a lower level of abstraction. This is because the abstract norms that the code of ethics inspires are illustrated by describing detailed cases and specific actions derived from the code's abstract norms.

Abstraction is expressed in this case also by the fact that situations are characterized not by their details but by the ethical issues they raise. This method of exposition is selected to help software engineers identify situations as specific instances of more abstract cases for which the code of ethics outlines norms that are accepted by the community of software engineers. Such identification may guide software engineers to act according to the code of ethics.

TASKS

1. Why is the concept of ethics abstract?
2. Tell a story related to software engineering that raises an ethical dilemma. Based on the Code of Ethics of Software Engineering suggest possible solutions. Tell another story whose solution can be derived from the solutions you offered for the first story. By identifying what the two stories have in common, explain why these similar features are sufficient for deriving the same solutions.

Chapter 9, "Program Comprehension, Code Inspections, and Refactoring," and Chapter 10, "Learning Processes in Software Engineering," focus on the team and some of its activities during the course of software development. Chapter 9 reviews the topic of program comprehension. This chapter highlights the cognitive aspect of software engineering by presenting different theories of program comprehension and examining code review processes. Abstraction plays a central role in this chapter, as some program comprehension theories use abstraction as their organizing idea. The relevance of abstraction in this context of program comprehension is clear. In the process of program comprehension, one has to move between levels of abstraction. This requires, of course, a lot of awareness on the part of the person who comprehends the computer program.

TASK

In what sense does awareness of the existence of different levels of abstraction improve program comprehension processes?

Chapter 10 focuses on learning processes. Specifically, two topics are addressed. The first topic is software engineering as a reflective practice. With respect to this topic, it is illustrated how the construction of a ladder of reflection may increase one's awareness of the existence of different levels of abstraction and, consequently, to increase the complexity of the objects with which one thinks. In other words, the discussion about software engineering as a reflective practice illustrates how one may increase the level of abstraction of one's thinking when reflection is interwoven in the process of software development.

The second topic discussed in Chapter 10 is learning organization. Specifically, Senge's framework for learning organizations [Senge94] was described. According to Senge, one of the dimensions of a learning organization is systems thinking, which helps us see more effectively how to change systems and act more in tune with the larger processes of the natural and economic world around us. In other words, systems thinking is a conceptual framework based on patterns that explain different events as instances of one phenomenon. We have said in that context that systems thinking is about abstraction, since it guides us to ignore details and to examine what different events have in common.

TASK

Suggest examples from the world of software engineering that illustrate the idea of systems thinking. Explain connections to the idea of abstraction.

We believe that at this stage readers gain quite a comprehensive picture about how abstraction can be seen in almost every phenomenon related to software engineering. Readers are invited to continue analyzing in a similar manner the next chapters of the book.

ADDITIONAL TOPICS RELATED TO ABSTRACTION

We end this chapter by discussing several topics that may expand readers' perspective on the concept of abstraction. Specifically, the following topics are discussed: the human aspects of software architecture, architecture versus design, quality attribute workshops, metaphors in science and philosophy, and abstraction in computer science and software engineering education.

The Human Aspects of Software Architecture

Software architecture is becoming a ubiquitous part of computing. It has nearly the same characteristics as design, although many of its practitioners often use "Software Architect" on their business cards, the same way "Software Engineer" appeared on the business cards of an earlier generation.

There is a functional side to software architecture. This side is well understood and easy to work with. It usually drives the shape of the software. There is another side, which the Software Engineering Institute (SEI) calls the "quality attributes." These are characteristics like usability, modifiability, safety, and similar "ilities." These are usually what the human user of an architecture cares about and are a subject of this section [Bass03]. We will also discuss the use of a metaphor and its contribution to architecture.

TASKS

1. What is a quality attribute?
2. How are quality attributes found?
3. How is a metaphor used to derive architecture?

Architecture is becoming an increasingly popular way to describe the shape of software. This shape frames development strategies. Since it involves the user, it is basic to software engineering.

Architecture versus Design

We have seen that "designing" is one of the verbs in the paradigm of software development. The activities associated with developing an architecture, as well as a design, are subsumed here. So, is there a difference between architecture and design? There is, just as there is a difference between high-level design and detailed design, again, both subsumed in the verb "designing." Detailed design shows how certain functionality is implemented, so it is closer to "coding." High-level design is the last chance to put the requirements of the "specifying" step out for validation. Therefore, it is closer to specifying.

Software architecture is a way of expressing high-level design that accounts for the quality attributes as well as the functionality of the software. People try to find architectures because if they can define a good one, later changes to the software will not cause a change in the architecture. This is a reason for prototyping, captured in building architecture by models and in Computer-Aided Designs (CAD). Building architects are adamant about prototyping because they can only erase defects is with jackhammers, or they can live with them, both expensive propositions.

Designs are fairly idiosyncratic; they do not carry the generalization of architecture. An architecture often hearkens back to the form of a successful high-level design. For example, in the architecture of buildings, the form of "gothic" has proven to be successful as part of a church, and many churches are in the gothic style, or have some gothic themes. In software, pipe-and-filter, blackboard, publish-and-subscribe, and other architectures provide a basis for detailed design. One way they do this is by identifying components and by incorporating quality attributes [Garlan and Shaw96].

Software architecture is a way of describing the entire detailed design. How are parts of the architecture found? One way is to identify quality attributes and prioritize them. The Unified Process (UP) claims that it is "architecture-centric" [Jacobsen99]. Agile processes, although they make no such claim, do almost the same. The well-known white board with the design written on it is a common part of agile workplaces [Beck00].

Quality Attribute Workshops

Many people think that agilism ignores architecture. There is no explicit architectural step, so it seems as if the architecture "falls out" in the flurry of coding. Actually, XP has at least four practices that contribute to an architecture. It has at least one major value that contributes to capturing the architecture. The SEI has at least four architecture evaluation methods that can apply to XP or agile methods in general.

The four XP practices that help engineers build the architecture are Customer On-Site, the Metaphor, Simple Design, and Refactoring [Beck00]. The customer is expected to not only provide the requirements, but to prioritize them (see Chapter 13). Each cycle of coding (about two weeks long) provides business value if produced according to the correct priorities. The difficulty lies in choosing these priorities. An experienced customer has an almost intuitive grasp of this. Others would benefit from a Quality Attribute Workshop (QAW) [Barbacci03]. This workshop would help customers prioritize requirements along the lines of their business needs, and would help identify the stories.

The QAW also has as a product some test cases to exercise the requirements developed. These can overcome one of the problems of customer involvement in XP. It is expected that customers develop user stories for requirements and then work on acceptance test cases for the end of development. Many customers do not know how to build these test cases. The QAW gives them clues and encourages them to build the cases. Moreover, building test cases fits in with the XP practice of "test-first" or "build for the test" philosophy, sometimes called "test-driven development," since test cases are available to test whether the code implements the requirements from early in the development ("acceptance tests"). These test cases can be built as code is being built, so the product of the software development team can be checked at the end.

In addition, the QAW method has as its main focus the development of "scenarios." These can be used to derive stories and as input to other SEI methods of revealing quality attributes.

The XP practice that affects software architecture the most is the Metaphor. The Metaphor is intended as a bridge between the technically oriented developers and the nontechnical customers. It is also to be used among developers. In both cases, it is intended to fulfill the value of communication. The Metaphor is meant to be a natural language description of the product. It is also meant to inspire the architecture [Beck00]. Metaphors are popular in philosophic thought, but are hardly limited to philosophy. Science and, now, software engineering make use of metaphors. They make much of abstract thinking achievable [Lakoff99].

The use of a metaphor in agile programming, especially XP, has two purposes: it is a start on the architecture of the system, and it is a means of communicating about the system with the user. For example, the first XP team visualized a payroll system as an "assembly line." Both clients and developers could discuss an assembly line, especially since they worked for an automobile manufacturer. Stations became functions. The metaphor sounded like it fit the "pipe and filter" architecture, of which the developers were aware (see Figure 11.1).

Both groups could discuss the assembly line metaphor. The technical team could use the architecture as they wanted, and keep the details from the user.

It seems that the metaphor, despite its success on the automobile company's payroll project, is clearly the least used of the XP practices [Herbsleb03]. Many more detailed descriptions of XP ignore it all together or replace it with "Stand-Up Meetings," a practice from Scrum (another agile method) [Anderson04]. At one point in 2002, Kent Beck himself (the originator of XP) said he was going to take an

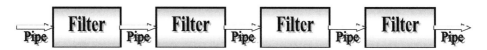

FIGURE 11.1 A pipe-and-filter architecture.

informal poll whether developers were using metaphors at that year's object conference. He said he would never speak of metaphors again, and would drop metaphors from XP if a majority was not using them.

Our research, documented in [Herbsleb03] and [Tomayko03], first aimed at finding some quantifiable reason to eliminate metaphors. We found that metaphors, even naïve ones, are rarely used. (A naïve metaphor is one in which it is close to the actual description of the software, as in controllers for stations on an assembly line and an assembly line.) The traditional poor communication about the structure of the software persists between client and developer. In the continuation of this chapter, we add a short section about metaphors in science and philosophy.

Another evaluation method that fits early in an iteration is the Cost/Benefit Analysis Method (CBAM) [Kazman02]. This method can help establish priorities for an Architectural Tradeoff Analysis Method (ATAM).[2] Since the ATAM is done to ensure that priorities are being met, the SAAM (an older method described in *Software Architecture in Practice* [Bass98]), or a CBAM and a QAW early on are sufficient. However, if an ATAM is used, then it is be factored into the end of first cycle, before the second cycle begins. The software architecture is pretty well set in XP by then. If the "XP customer on site" practice is being followed, most of the stakeholders (required to be present in SEI methods) are again easy to find to participate. An ATAM at this time would reveal if the results of the QAW were still being followed, as well as if the nascent architecture embodied other priorities. If not, it is early enough for inexpensive and easier prioritizing of stories and refactoring.

Now we consider the question of documentation. Many believe that XP rejects documentation. This is not true. The Agile Manifesto [Martin02] says that Agilists favor human interaction over extensive documentation. In short, documentation is not ignored; it is put off until the documented item is stable, with few changes expected. The documentation for the architecture in XP, then, is most often on a white board during the cycles. At the end of development, an XP team can document the architecture, using [Clements02] as a guide. If the team is to disband and another takes over, documentation is doubly important. The best guideline for the product is that the code bears the burden of most documentation. Any additional documentation is to assist in understanding the code. This is a good time to see if the product could meet product line guidelines as established in [Clements01]. If so, some refactoring should be done to provide interfaces.

TASK

Does the time involved in doing an SEI method preclude its use in agile approaches to software development?

Metaphors in Science and Philosophy

There is an extensive prior literature on metaphors, since they are frequently used in philosophy and science [Brown03, Jones82, Lakoff99, and Ortony79]. In [Brown03], the author explains how scientists work, the nature of scientific knowledge, and an introduction to metaphors in general. Then he describes in some detail "conceptual metaphors." The general notation for metaphors of this type is "Target domain is source domain." Take the sentence:

"She can *stonewall* his opinion."

Hence, "Argument is construction." This is a conceptual metaphor.

For years, scientific principles were represented as conceptual metaphors (see also [Jones82]). Perhaps the most famous is the atom as a solar system, which lasted until the bewildering variety of subatomic particles were theorized or discovered and shown to have components themselves.

[Lakoff99] refers to "primary metaphors." There are a number of examples, like "more is up." In other words, increasing quantity is shown by metaphors containing "up" as in "prices are high." This idea develops in childhood. There are neural connections that develop to make this type of metaphor very powerful. For example, the primary metaphor "Bad Is Stinky" underlies "Code Smells," leading us to refactoring. Simple design and Refactoring are frequently found as patterns of code. Ideas for changing these are found in [Gamma95 and Fowler99].

TASK

Develop a metaphor and a resulting architecture for an automatic teller machine (ATM). Is this an example of a naïve metaphor? Why, or why not?

Abstraction in Computer Science and Software Engineering Education

At the *Sixteenth Conference of Software Engineering Education and Training*, held in 2003 in Madrid, Spain, Jeff Kramer of Imperial College, London, presented his keynote talk: "Abstraction—is it teachable? Or, the devil is in the details." One of his main messages was that abstraction is not a simple concept to teach (or learn!)

and that the idea of abstraction should be addressed throughout the entire software-engineering curriculum.

In what follows, we illustrate the central role that abstraction plays in computer science and software engineering programs by describing the recommendations of three committees that suggested structures for computer science and software-engineering undergraduate programs. In 1991 The Computing Curricula 1991 Committee[3] identified the following three concepts as ideas that each computer science curriculum should emphasize: theory, abstraction, and design. This recommendation is explained by the fact that mastery of the discipline includes not only an understanding of basic subject matter, but also an understanding of the aforementioned three points of view that computing professionals employ and students need to appreciate.

In a similar spirit, 10 years later, the 2001 report of the Computing Curricula 2001 Committee[4] explains that one of the principles that guided the work of the committee was the fact that the foundations of computer science are drawn from a wide variety of disciplines, and consequently, undergraduate study of computer science should require students to use concepts from many different fields. Accordingly, computer science students should learn to integrate theory and practice, to recognize the importance of abstraction, and to be able to appreciate the value of good engineering design.

The same orientation is also expressed in the Computing Curriculum—Software Engineering (CCSE) Public Draft[5] published on July 17, 2003. Its Principles section describes the foundational ideas that guided the development of the CCSE materials. The second principle states that all software-engineering students must learn to integrate theory and practice and to recognize the importance of abstraction and modeling.

As can be observed, the same principle guided the work of all committees. In fact, we tried to convey a similar message in this chapter when we addressed, from the perspective of abstraction, topics that were discussed previously in the book.

SUMMARY QUESTIONS

1. Summarize the main ideas of this chapter. Can you describe them in terms of different levels of abstraction?
2. How is abstraction reflected in the structure of this book (described in the introduction of the book)?
3. Find information about the evolution of programming languages. Explain connections between different characteristics of each language and the

heuristics discussed in this chapter. Explain how each generation of programming languages helped reduce some of the cognitive complexity involved in software development.

REFERENCES

[Abelson and Sussman86] Abelson, Harold, and Sussman, Gerald J., *Structure and Interpretation of Computer Programs*, MIT Press and McGraw-Hill, 1986.

[Anderson04] Anderson, David J., *Agile Management for Software Engineering*, Prentice Hall, 2004.

[Auer02] Auer, Ken, and Miller, Roy, *Extreme Programming Applied*, Addison-Wesley, 2002.

[Barbacci03] Barbacci, Mario R.; Lattanze, Anthony J.; Stafford, Judith A.; Weinstock, Charles B.; and Wood, William G., *Quality Attribute Workshops (QAWs)*, Third Edition, CMU/SEI-2003-TR-016, 2003.

[Bass98] Bass, Len; Clements, Paul; and Kazman, Rick, *Software Architecture in Practice*, Boston, MA 1999.

[Bass03] Bass, Len; Clements, Paul; and Kazman, Rick, *Software Architecture in Practice*, Second Edition, Boston, MA 2003.

[Beck00] Beck, Kent, *Extreme Programming Explained*, Addison-Wesley, 2000.

[Brown03] Brown, Theodore L., *Making Truth: Metaphor in Science*, Illinois University Press, 2003.

[Clements01] Clements, Paul, and Northrop, Linda, *Software Product Lines: Practices and Patterns*, Boston, MA, 2001.

[Clements02] Clements, Paul; Bachmann, Felix; Bass, Len; Garlan, David; Ivers, James; Little, Reed; Nord, Robert; and Stafford, Judith, *Documenting Software Architectures: Views and Beyond*, Boston, MA, 2002.

[Dijkstra72] Dijkstra, Edsger, W. "Notes on Structured Programming." In Dahl, O. J.; Hoare, C. A. R.; and Dijkstra, E. W. (eds.). *Structured Programming*, Academic Press, 1972.

[Fowler99] Fowler, Martin, *Refactoring*, Boston, MA, 1999.

[Garlan and Shaw96] Garlan David, and Shaw, Mary, *Software Architecture: Perspectives on an Emerging Discipline*, Prentice Hall, 1996.

[Gamma95] Gamma, Erich; Helm, Richard; Johnson, Ralph; and Vlissides, John M., *Design Patterns: Elements of Reusable Object-Oriented Software*, Addison-Wesley, 1995.

[Herbsleb03] Herbsleb, James; Root, David; and Tomayko, James, *The eXtreme Programming (XP) Metaphor and Software Architecture* (CMU-CS-03-167), School of Computer Science, Carnegie-Mellon University, 2003.

[Hoare86] Hoare, C. A. R., "Mathematics of Programming." *Byte* (August 1986): pp. 115–124, 148–150.

[Jacobsen99] Jacobsen, Ivar; Booch, Grady; and Rumbaugh, James, *The Unified Software Development Process*, Addison-Wesley, Boston, MA, 1999.

[Jones82] Jones, Roger S., *Physics As Metaphor*, University of Minnesota Press, 1982.

[Kazman02] Kazman, Rick; Asundi, Jai; and Klein, Mark, *Making Architecture Design Decisions: An Economic Approach*, CMU/SEI-2002-TR-035, 2002.

[Kramer03] Kramer, Jeff, Abstraction—is it teachable? Or the devil is in the detail, keynote at the *Sixteenth Conference of Software Engineering Education and Training*, IEEE Computer Society, Madrid, Spain, 2003.

[Lakoff99] Lakoff, George, and Johnson, Mark, *Philosophy in the Flesh*, Basic Books, 1999.

[Martin02] Martin, Robert, and Newkirk, James, *Extreme Programming in Practice*, Addison-Wesley, Boston, MA, 2002.

[Ortony79] Ortony, Andrew, ed., *Metaphor and Thought*, Cambridge University Press, 1979.

[Pattis81] Pattis, Richard, *Karel the Robot*, John Wiley & Sons, Inc., 1981.

[Senge94] Senge, Peter M., *The Fifth Discipline: Fieldbook*, A Currency book, published by Doubleday, 1994.

[Tomayko03] Tomayko, James E., and Herbsleb, James. *How Useful Is the Metaphor Component of Agile Methods? A Preliminary Study* (CMU-CS-03-152). School of Computer Science, Carnegie-Mellon University, 2003.

[Wirth71] Wirth, Niklaus, "Program Development by Stepwise Refinement." *Communications of the ACM* 14(4), (1971): pp. 221–227.

ENDNOTES

[1] It is stated explicitly in the Code of Ethics of Software Engineering that "[t]his Code may be published without permission as long as it is not changed in any way and it carries the copyright notice." These two requests are fulfilled in this book.

[2] For additional information about these concepts, go to the Software Engineering Institute Web site at *http://www.sei.cmu.edu*.

[3] The ACM/IEEE-CS Joint Curriculum Task Force, Computing Curricula 1991 Report: *http://www.computer.org/education/cc1991/eab1.html*.

[4] Computing Curricula 2001 Report: *http://www.computer.org/education/cc2001/final/index.htm*
[5] Computing Curriculum—Software Engineering (Draft of July 17, 2003): *http://sites.computer.org/ccse/volume/FirstDraft.pdf*

12 The Characteristics of Software and the Human Aspects of Software Engineering

In This Chapter

- Introduction
- Objectives
- Study Questions
- Relevance for Software Engineering
- Software Characteristics
- Programming Style
- Evaluation of Programming Style
- Affective Aspects of Human-Software Interaction
- Summary Questions

INTRODUCTION

While reading this book, one can observe that software characteristics can be examined from different perspectives. For example, Chapter 4, "Software as a Product," looks at software from the customer's perspective; Chapter 9, "Program Comprehension, Code Inspections, and Refactoring," deals with program comprehension and examines software characteristics from the developers' perspective. This chapter continues with the examination of software characteristics from the developers' perspective.

This kind of examination is derived from the process-oriented perspective toward software development presented in Chapter 7, "Different Perspectives of Software Engineering." Continuing the process-oriented perspective, this chapter illustrates that even software characteristics that seem to be connected only to the software itself cannot be isolated and detached from the software developers. Specifically, we focus on communication issues related to programming style. In addition, we find it appropriate to discuss in this chapter the developer-software relationship from the affective perspective by focusing on the activity of debugging, which is disliked by many developers.

OBJECTIVES

- Readers will be able to identify what is required from software engineers when they are asked to produce software with a specific characteristic.
- Readers will be able to suggest specific development practices to ensure that the software they produce meets its specifications, not only from the customer's perspective, but also as a qualified product.
- Readers will conceive of programming style as a communication means between software developers.
- Readers will increase their awareness of affective aspects of software development.

STUDY QUESTIONS

1. Name two positive and two negative characteristics of software. Explain your choices. In what way do these characteristics influence the daily life of software developers?
2. Review several software-engineering books. List the main properties of software described in these books. Explain what is required from software developers to produce a software system that is characterized with each of the properties you listed.
3. Search the Internet for "software characteristics." Review several Web sites and list the main software characteristics you find. Analyze these characteristics according to their influence on the daily work of software engineers.
4. For each characteristic you mentioned in Questions 1 through 3, explain how one can determine whether that property characterizes a specific software system.

5. For each characteristic you listed in Questions 1 through 3, check its relevance to different types of software tools (educational systems, business tools, and computer games, for example). Conclude: Are different characteristics relevant for the description of different types of software tools or, alternatively, are the same characteristics relevant for the description of all types of software tools?
6. Chapter 2, "Software Engineering Methods," outlines in detail three software development methods: Spiral Model, Unified Process, and eXtreme Programming (XP). Analyze each of these methods according to the software characteristics that guides its production.
7. What is your most liked development activity? What is the your most disliked development activity? Why? Do your colleagues have similar feelings?

RELEVANCE FOR SOFTWARE ENGINEERING

The object we look at in this chapter is the software itself. It has two facets: the thing that is constructed—the code itself—and the product that the code generates when the code is executed. The first facet is closely related to the software developers; the second is related mainly to the customer. The latter has been already discussed in Chapter 4. This chapter looks at the code itself and examines it as an object.

One may wonder why it is important to examine the code at all. After all, we develop software tools to satisfy our customer's needs. Accordingly, as far as the customer uses the product satisfactorily, we should not be bothered by the code at all. This argument is incorrect for at least three reasons.

First, from the process-oriented perspective presented in Chapter 7, the development period continues after the software is shipped to the customer. This observation is clear to us if it is examined from the customer's perspective as well. Only when customers start using a software system can they get real opportunities to understand what benefits the software can provide them. This understanding may lead the customers to ask for new features or to change existing features. Thus, software systems are in a development process throughout their entire life cycle. This reality should not be ignored from the first day of the software development. Accordingly, developers should conceive of the code as an object that will be reshaped for a long period of time.

Second, if a software system is not written according to some basic programming principles, it simply cannot be developed. There are cases in which the code is written so messily that it is impossible to understand either its structure or its logic. Consequently, no further development can be carried out. In such cases, it is better to throw the code away and start developing new code from scratch.

Third, there are clear connections between the characteristics of the code and the quality of the software that the customer gets. For example, it is important to enable customers to ask for new features after they start using the software system. If the software is written in such a way that does not allow the addition of new features and customers need new features in the future, from the customers' perspective the software is developed in an unprofessional manner.

There are connections between the aforementioned explanations and the topic discussed in Chapter 2. However, it is not our intention to evaluate the code that is produced by each software development method. No matter what software development method one uses, the produced code should have some basic characteristics that ensure a qualified product both from the customer and developer perspectives.

SOFTWARE CHARACTERISTICS

Although it was mentioned earlier in this chapter, we want to emphasize the scope of our discussion. We talk about software characteristics—the characteristics of the code. We do not address characteristics that refer to what the code creates when it is executed. Such characteristics are usually expressed by properties such as usefulness, user friendly, graphical user interface (GUI), and so forth.

Code characteristics can be divided into two types: characteristics that examine the functionality of the code (efficiency, maturity, testability, and recoverability, for example) and characteristics that are concerned with the code's shape and communicative aspects (readability, clarity, meaning). See [Shneiderman80] for further information.

The discussion in this chapter is limited to the second type of characteristic, since these characteristics are most relevant to the human aspects of software engineering described in this book. The awareness of such characteristics invites software developers not only to consider their current needs to produce the code, but also, to analyze other team members' perspective and understanding of the code, and to consider topics related to the future development process. We discuss how such characteristics are achieved by referring to programming style.

The following discussion about programming style is connected neither to a specific programming language nor to a specific software development environment. Our aim is to show how programming style may be a vehicle that improves communication between software developers. For this aim, we look only at programming style guidelines that address communication issues and not technical issues. For example, using global variables is a bad habit, because their value can be changed by many programmers and no one can be sure what their value is in any specific moment. Still, no communication problem between software developers will result from using such variables. Accordingly, because this is a more technical problem than a communication among programmers problem, we do not address it.

PROGRAMMING STYLE

The aforementioned desired software characteristics may be achieved by applying appropriate programming style. This topic is tightly connected to Chapter 9, which addresses program comprehension. In what follows, we strengthen this connection by illustrating how programming style may influence directly and implicitly program comprehension.

TASKS

1. Identify at least three connections between programming style and program comprehension. Illustrate these connections by short stories that may happen in the daily life of software engineers.
2. There are specific programming style guidelines for specific programming languages (such as C++ and Java). Find such guidelines (several are available on the Web). What aspects of the code do these guidelines address? Select five rules from one of these guidelines and explain their connection to program comprehension.
3. Identify additional chapters of this book whose topics may have direct connection to programming style.
4. Look for information about the "Y2K bug." What issues related to programming style are connected to this bug?

The most important feature of communicative programming style is that the code communicates itself without the need for comments, explanations, apologies, and so forth. For example, the essence of a variable can be expressed by its name.[1] If one names a variable accordingly, one does not have to add a comment that explains the variable's nature and role. Accordingly, when you write a computer program, think about the other software developers who will have to work with the code, as readers of a book. In this sense, a comment is analogous to a footnote, which sometimes interrupts a reader's fluent reading. In book writing, however, there are cases when it is more logical to include endnotes and clarifications; there are cases in programming when comments should be added.

In addition, for the developer of the code, the addition of comments is not a good habit when it is applied in too many situations (as on each line of assembly code). Think about the habit to document in detail any small piece of code. If in later stages one wants to change one's code (see refactoring in Chapter 9), one would probably balk at the extra documentation work needed if changes are introduced into the code. Recording these thoughts may result in not improving the code.

Let us assume that this extra effort does not bother you and you do change the program when you find opportunities to improve it. In many cases, programmers forget to update the documentation in a way that corresponds to the updated code. Consequently, we may end up with documentation that does not fit the code that it documents. In such cases, think about the programmers who will have to work with code that is wrongly documented. Clearly, computer programs should be written in a way that enables future developers who will work with the code to perform their work easily.

TASK

Write two computer programs that execute the same task. The programming style of the first one requires the addition of many comments in order to understand it. The second computer program is written in a way that no comment (or almost no comments) is needed for its comprehension.

a. Give the two programs to two programmers and ask them to explain the program they received. Observe and document the process through which both of them go. Draw your conclusions.
b. Ask each of the two programmers to make the same modification in the program. Trace the change process in each case. What are your conclusions?

There are many specific programming guidelines related to programming style. Instead of just listing them all, we suggest using three mechanisms that may support the application of a communicative programming style: abstraction (see Chapter 11, "Abstraction and Other Heuristics of Software"), refactoring (see Chapter 9), and simplicity. Naturally, additional mechanisms may be employed to achieve good programming style.

Abstraction

Abstraction was discussed in length in Chapter 11. Here we describe how it may be applied to programming style. We start by examining the names we assign to variables, functions, and any other object for which we can define a name in a computer program.

TASK

Why do people have names? Why do objects have names? What communication benefits do we get from using names?

Your answers to the preceding task may give you the right framework in which to think about the names you assign to functions, variables, and other identifiers in your programs. Names are given to identifiers to improve communication among programmers who work on the same code, including the one who writes the code. Inappropriate naming will not enable the person who develops the code to understand it a short period of time after the development is completed

The names we give to different things should be meaningful and reflect the essence of what we name in a mnemonic way. Abstraction is relevant here, as it enables us to forget about the details of how things are constructed or how they operate, and to encapsulate their essence in their name.

The desire to capture in one name the meaning of a thing leads to some applicable rules. Here are two examples:

- Functions and procedures should be short enough so that they perform one task that can be reflected in their name.
- One-character identifiers are not a good choice for naming objects. This is because such identifiers may not be informative (unless it is a convention to use such identifiers as in mathematical expressions, loop variables, and array indices).

One case that illustrates the idea of expressing the essence of things by giving them appropriate names is that of constants and enumerated types. Enumerated types allow us to declare a variable with a limited range of values, which can be numeric or any arbitrary strings. Accordingly, if you want a variable to store the days of the week, instead of storing the integer values of 1, 2, 3, ... for Monday, Tuesday, Wednesday, it makes much more sense to give the enumerated values meaningful names.

Refactoring

The idea that computer programs are constructed in steps is clear. Here, we describe how programming style can be improved throughout a development process in which refactoring actions are intertwined (see Chapter 9).

Refactoring guides us not to stop working on a piece of code when we complete its first tested draft. Let us consider the book metaphor we used earlier. Authors do not stop working on a book as soon as they complete the first draft; rather, they keep improving it. For example, they look for better words to express the exact meaning of an idea they are trying to deliver. After these and many other small improvements, readers eventually get the feeling of reading a well-written book.

Another image taken from the world of book writing is the author, upset with what he has written throwing crumpled pages into the trash bin. The full-with-pages trashcan next to his table indicates that, instead of trying unsuccessfully to reformulate an idea he wants to communicate, he prefers to start writing from scratch on a white page. Similarly, there are cases in code developing in which it is not worth patching bad code, but rather it is better to rewrite it.

TASK

Suggest additional analogies between the process of book writing and the process of developing software systems in general and refactoring in particular. In what ways do these analogies help in understanding the nature of these two processes?

Simplicity

Let us continue with the metaphor of book writing and reading. It is reasonable to assume that when one reads a book, one would prefer that the book be written clearly, even if it delivers complicated ideas. This should also be the nature of the way in which software code is developed. The code should be simple, direct, and clear. Indeed, some programmers need reminding that clarity should not be sacrificed for efficiency. Since this efficiency is usually obtained by difficult and tricky expressions, in many cases the small gains that this efficiency adds are lost eventually in programmers' misunderstandings and confusions.

TASKS

1. Suggest at least three guidelines that may help software engineers gain simplicity in the code they produce.
2. How does XP (Chapter 2) treat the value of simplicity?

EVALUATION OF PROGRAMMING STYLE

This chapter looks at some programming guidelines that may have direct influence on program comprehension and on the communication between software developers who work on the program in different stages of its life cycle. How can the influence of these guidelines on the developers' work be measured? It is difficult to measure code characteristics of the kind discussed here with quantitative evaluation methods; qualitative methods (see Chapter 4) are more appropriate in this case.

Since most software engineering books discuss at length the topic of software evaluation and measurement, our discussion of this topic is brief. We conclude with one lesson worth remembering in any evaluation process (not only of software): "Measure what is important to measure, *not* what is easy to measure."

AFFECTIVE ASPECTS OF HUMAN-SOFTWARE INTERACTION

In this section, we analyze the observation that debugging is one of the most disliked activities of any software developer. Specifically, we analyze this phenomenon from the perspective of freshmen majoring in computer science at the end of the second semester of their study. Freshmen dislike the debugging process. Six explanations for this phenomenon are suggested. In addition, the activity of debugging as a learning activity is discussed.

TASK

Speculate: Why is debugging disliked?

We decided to include this observation in this chapter because debugging is mentioned by freshmen computer science students as one of the most disliked topics in computer science. Since debugging is an activity both computer science students and practitioners deal with, it is important to understand why both feel this way and to find ways to change this feeling. Naturally, this conflict is about software characteristics and human interaction with software and, as such, is located in this chapter. As in Chapter 3, "Working in Teams," we decided to include a university phenomenon since we hope that students will be one of the audiences of the book.

The data for this section was collected by a questionnaire distributed to a group of 71 freshmen (15 women and 56 men) at the end of their first year of study learning the CS2 (introduction to systems programming) course (see the appendix to this chapter). The majority of the students (14 of the women and 53 of the men) were in their second semester of study and had taken CS1 (introduction to computer science for computer science majors) in the previous semester. The questionnaire was set up with the aim of learning about freshmen's conception of computer science at the end of their first year. The decision to give the students the questionnaire at the end of the first year was based on the assumption that although the students did have some idea of computer science when they entered the university, at the end of the first year of study this picture was reshaped, and that at the end of the first year of study the students are less influenced by pre-university factors. We asked students how they perceive the discipline, what they like and dislike in computer science, as well as other related questions about the discipline in general and about basic concepts of the field in particular.

CS1 and CS2 usually address the basics of computer science as they are conceived by the program orientation in general and instructor preferences in particular. In most cases, these ideas are expressed through the teaching of programming by using some programming language. In the case of those students whose responses are discussed here, CS1 focuses on algorithms and complexity and is taught using the C programming language. No prior knowledge in computing is required for that course. According to the course Web site, the CS2 course aims to advance the student to the level of large-scale software systems programming. Special emphasis is given on software design techniques and tools.

The instructor who taught this particular CS2 course is a commendable instructor, who usually receives high student evaluation scores and has a great deal of industrial experience.[2] The lectures addressed much more than just programming; theoretical topics, such as abstract data types, and human-oriented topics, such as the customer point of view, were included.

In what follows, we focus on the observation, mentioned previously, that is derived from the analysis of the questionnaires. According to this observation, students are discouraged by the debugging process. The dominance of debugging as a negative "thing" is even highlighted when we put it together with the dominance of "orderly way of working" and "(early) design" that were presented by the students as activities that should be watched in software development.

This observation is derived from students' responses to questions (e) and (f) (see Appendix). As can be observed from Table 12.1,[3] the predominant answer to question (e) is *debugging*; the predominant answers to question (f) are *(early) design* and *orderly way of working*. As explained later, this observation is supported from different angles by students' answers to several of the other questions that appear in the questionnaire.

TABLE 12.1 Students' Answers to Questions (e) and (f)

(e) What are the three "things" that you dislike about computer science?
Debugging (21)
Mathematics (9)

(f) In your opinion, what are the three main principles that should be watched in software development?
(early) Design (20)
Orderly way of working (19)
Efficiency (12)
Consideration of future versions and maintenance (13)
Simple code (12)
Understanding the problem (11)
Modularity (5)
Documentation (4)

In what follows, we present six explanations for this observation.

Conceptual perspective: This explanation refers to the special attention that debugging gets in students' answers. As can be observed in Table 12.2, although students conceive of computer science as more than just the computer (Question (a)), at the same time, programming (of algorithms) is conceived of as the main topic of computer science (Question (b)), and coding is one of the main activities that people working in computer science carry out (Question (c)). Since in Questions (b) and (c) many students mentioned coding and other activities related to coding, it is clear that students conceive the programming aspect of computer science to be of more importance.

With such a focus on the activity of coding, debugging is mentioned as the third item on the computer scientists' activities list. The fact that it is conceived as so central in the life of computer scientists may explain the expression of any feeling toward it. The next explanations attempt to explain why these feelings toward debugging are negative.

TABLE 12.2 Students' Answers to Questions (a) through (c)

(a) In your opinion, what is "computer science?"
The computer (20)
Programming (19)
Anything that is connected to computers (13)

(b) In your opinion, what are the three main topics of computer science?
Algorithms/programming (58)
Computer/hardware (22)
Development methodologies (8)

(c) In your opinion, what are the three main activities that people working in computer science carry out?
Coding (38)
Maintenance/support (15)
Debugging (14)
Algorithm development (13)
The path: Design–coding–testing (10)
Management issues: Project management, meeting customer requirements, money (7)

Debugging is sporadic: Usually, students are taught to solve problems systematically, and the problems they are asked to solve have an algorithm for solving them. For example, in the context of mathematics education it is well known that students feel very strongly that mathematics always gives a rule to follow to solve problems they face [Carpenter83]. Naturally, these mathematical habits are brought by students to the act of programming. However, in contrast to many of the mathematical problems with which students had to deal, Francel and Rugaber [Francel and Rugaber01] argue that after a half century of computer program development, no "best" method for debugging programs is known. It is suggested that when students are unable to apply their habits of work to the debugging part of coding, they conceive of debugging as an irregular process, and tend to feel uncomfortable with it.

Debugging is associated with mistake finding: We would argue that since debugging is conceived as finding mistakes, it is natural that it raises resistance. A similar observation is observed among practitioners. For example, according to Cohen, Birkin, Garfield, and Webb [Cohen04], many developers view their code as an extension of themselves and thus take it personally when someone finds it fault with it. In their article, they report about testers who cite developers' reluctance to accept the existence of error. In general, no one likes that his or her mistakes are found, and when, in the case of students, they find their own mistakes, it is even worse.

This perception of debugging may be the result of the way in which many computer programs are presented to students in polished and complete form. Such a presentation may give students the wrong impression that no debugging process is interwoven here and that they are the only ones who make mistakes in software development. Because students are usually taught that mistakes are negative phenomena, it is reasonable to assume that they would have negative feelings toward a process that is based on mistake finding. This incorrect interpretation of bugs is addressed by Pattis who, after laying out in detail the development of a relatively long program in the *Karel the Robot* environment, advises the learners not to succumb to the "I've got to get it perfect the first time" syndrome, since it may lead to programmer's block [Pattis81].

Debugging is disruptive activity: In contrast to code writing (one of the loved activities—see Table 12.3), which usually leads students to the accomplishment of their tasks, the debugging process does not naturally lead one toward the achievement of his or her target. Furthermore, as some students have no previous experience with code writing, they may make relatively many programming mistakes that result in a long debugging process. Moreover, fixing one bug may introduce a new bug, and thus move one back to square one, so the level of uncertainty in the debugging process is increased.

Polya [Polya77] distinguishes four stages of problem solving. The first is based on the *understanding* of the problem. Second, we *plan* based on the observations of how the various items are connected and how the unknown is linked to the data. The third stage is *carrying out* the plan. At the fourth stage, we *look back* at the completed solution and review and discuss it. With respect to the fourth stage—looking back—Polya says that even fairly good students tend to skip it. In doing so, they miss an important and instructive phase of the work. Polya argues that by looking

TABLE 12.3 Students' Answers to Question (d)

> (d) What are the three "things" you like about computer science?
> Coding and algorithms (46)
> The queries (10)
> The computer (10)
> Conceptual aspects and design (6)

back at the completed solution, by reflecting on the path that led to it, learners can consolidate their knowledge and develop their ability to solve problems. As can be observed from Tables 12.2 and Table 12.3, the freshmen recognize the importance of Polya's first three stages—understanding the problem, planning the solution (design), and carrying it out (coding). However, they dislike the debugging process: not only do they have to look back (a habit they are not used to as Polya mentions) and not only do they have to find their own mistakes, but also there are times when they have to devote more time to this process than to finding, planning, and coding a solution to a problem. Thus, it is not surprising that students dislike debugging.

> **Debugging is not part of the curriculum:** An examination of books that review the entire process of software development reveals that such books usually do not mention debugging as part of the software life cycle. Rather, testing appears as one of the last stages of software development (requirements, specifications, design, coding, testing, operation, and maintenance). In other words, debugging is presented as an activity that should be interwoven into the phases of testing and maintenance, but does not stand on its own, although it does require a lot of time and effort. Being an activity that is not a recognized part of software development and one that may interfere with the completion of a computer program may even cause an increase in students' negative feelings toward debugging. Furthermore, because debugging is not taught explicitly in all books (or lectures), students may feel that they do not really know how to do it.
>
> **Social perspective:** This explanation relies on a social perspective that has emerged from the students' responses to the questionnaire. As it turns out, students are unfamiliar with software failure but are aware of the powerful in-

fluence of software systems on everyday life (see Table 12.4). As can been seen, students refer mainly to transportation, military, financial, and medical systems. One of the students declared: "From the crash of a computer to a world war." Although most students admitted that they are unfamiliar with specific software disasters, they could imagine the possible results of such a disaster. This fact is important in itself. We suggest that some insights with regard to software influence on real-life situations be added to the students' overview of computer science.

Since debugging has a direct relationship to software failure, software failure is another perspective from which it is interesting to explore students' dislike of debugging. More specifically, since students are not familiar with specific software failures, they may not appreciate the importance of debugging.

The preceding analysis leads us to suggest addressing the process of debugging as a learning activity. Specifically, the debugging process may be presented as a meta-process that accompanies all the activities associated with software development (from requirements analysis to maintenance). Thus, debugging may acquire its importance in the form of a habit that supports one's progress in software development in general.

Accordingly, we suggest encouraging students to reflect on the debugging process and letting them feel that debugging processes contribute to their learning. The call for a reflective mode of thinking was mentioned previously in this book in Chapter 10, "Learning Processes in Software Engineering."

TABLE 12.4 Students' Answers to Question (g)

(g)	Are you familiar with a disaster that occurred as a result of software failure? If yes, what was the disaster and what was the reason? If you are not familiar with a disaster that occurred as a result of software failure, can you think of any disasters that could occur as a result of software failure?

Airplanes (18)

Missiles (11)

Financial systems (8)

Y2K Bug (9)

Atomic weapon (5)

Medical systems (hospitals) (5)

As explained in Chapter 10, this perspective is oriented from Schön's Reflective Practitioner perspective ([Schön83] [Schön87]). In general, Schön analyzes the added advantages of continuously examining one's practice and one's thinking about that practice. It is suggested that by well-designed activities, students may be taught to become reflective practitioners in general and use these skills to improve their debugging processes in particular. Some related ideas are presented in [Hazzan02].

It is interesting to note that in practice, developers' resistance to debugging processes has already been identified. Consequently, to overcome this resistance, agile software development methods in general and XP in particular introduced the practice of test-driven development [Kent03]. According to the practice of test-driven development, developers write *automatic tests* before they write the code. In this way, the debugging process becomes a game in which one's object is to pass all tests that one has set. The game is not annoying since the tests are automatic. This approach fits very well Hamlet and Maybee's detective approach toward testing [Hamlet and Maybee01]. They explain that testing is detective work that would be a serious challenge for Sherlock Holmes. From this perspective, the program and its specification and design document contain the clues to the crime, and the tester's role is to find the bugs. They admit, however, that not everyone likes detective work and that many people may not have the talent for it (p. 397).

SUMMARY QUESTIONS

1. Simple design (one of the eXtreme Programming practices) is a software characteristic.
   - What is required from software developers to produce a code according to this practice?
   - Discuss connections between simple design and refactoring.
   - How do these two practices support the XP values of communication and simplicity?
2. List software characteristics of different kinds. Sort them according to the level of accuracy according to which they can be measured.

FOR FURTHER REVIEW

1. Select one or more qualitative research tools described in Chapter 4. Design a small-scale study that examines the influence of specific programming style guidelines on how programmers manage to develop a computer program. Conduct the research and describe your conclusions.

2. Find out about the measurement of quantitative features of computer programs. Analyze connections between these measurements and the ideas of programming styles described in this chapter.
3. According to the measurements you discussed in Question 2, evaluate the last two computer programs you wrote. What conclusions can you derive from these measurements?
4. Interview some of your colleagues about their feelings toward debugging. Record their feelings and try to interpret them. Can you explain the source of these feelings?

REFERENCES AND ADDITIONAL RESOURCES

[Carpenter83] Carpenter, T. P.; Lindquist, M. M.; Matthews, W.; and Silver, E. A.; "Results of the Third NAEP Mathematics Assessment: Secondary School," *Mathematics Teacher* 76 (1983): pp. 652– 659.

[Cohen04] Cohen, Cynthia F.; Birkin, Stanley J.; Garfield, Monica J.; and Webb, Harold, W., "Managing Conflicts in Software Testing," *The Communications of the ACM* 47(1), (2004): pp. 76–81.

[Francel and Rugaber01] Francel, Margaret, A., and Rugaber, Spencer, "The Value of Slicing While Debugging," *Science of Computer Programming* 40 (2001): pp. 151–169.

[Hamlet and Maybee01] Hamlet, Dick, and Maybee, Joe, *The Engineering of Software*, Addison-Wesley, Inc., 2001.

[Hazzan02] Hazzan, Orit, "The Reflective Practitioner Perspective in Software Engineering Education," *The Journal of Systems and Software* 63(3), (2002): pp. 161–171.

[Kent03] Kent, Beck, *Test-Driven Development: By Example*, Addison-Wesley, 2003.

[Pattis81] Pattis, Richard E., *Karel the Robot*, John Wiley & Sons, 1981.

[Polya77] Polya, George, *How to Solve It?* Princeton University Press, 1973.

[Schön83] Schön, Donald A., *The Reflective Practitioner*, BasicBooks, 1983.

[Schön87] Schön, Donald A., *Educating the Reflective Practitioner: Towards a New Design for Teaching and Learning in The Profession*, Jossey-Bass, 1987.

[Shneiderman80] Shneiderman, Ben, *Software Psychology—Human Factors in Computer and Information Systems*, Winthrop Publishers, Inc., 1980.

Brian, Kernighan, and Plauger, P. J., *The Elements of Programming Style*, Second Edition, McGraw-Hill, 1988.

Conway, Damian, CSE2305/CSC2050 Course: Object-Oriented Software Engineering, School of Computer Science and Software Engineering, Monash University, Topic 14: Software Characteristics and Metrics: *http://www.csse.monash.edu.au/~damian/CSC2050/Topics/07.14.SWEng2/html/text.html#software_product_characteristics*.

PROGRAMMING STYLE LINKS

- **C++ Programming Style Guidelines:** *http://geosoft.no/style.html*
- **Java Programming Style Guidelines:** *http://geosoft.no/javastyle.html*
- **Lisp Programming Style:** *http://www.elwoodcorp.com/alu/table/style.htm*
- *http://ei.cs.vt.edu/~cs2604/Standards/Standards.html*
- *http://users.erols.com/blilly/programming/The_Elements_of_Programming_Style.html*
- *www.eecs.harvard.edu/~ellard/CS50-95/programming-style.html*

ENDNOTES

[1] According to the Confucian doctrine of "Rectification of Names," such names must be meaningful. Confucius was the first programmer!

[2] We would like to thank Dr. Yechiel Kimchi from the Department of Computer Science at the Technion for his cooperation.

[3] The table presents the answers that students mentioned more frequently. Numbers in parentheses indicate the number of students who mentioned each item.

APPENDIX—QUESTIONNAIRE

Note: The questionnaire appears here in a condensed format. It was given to the students on two pages with spaces between the questions for students' answers.

Name (not compulsory): _____

Semester of study: _____ Department: _____

Gender: F / M

In what semester did you learn CS1?
Winter 02   Spring 01   Winter 01   Before

In the following answers, you are requested to express your personal opinion according to your experience with computer science so far.

(a) In your opinion, what is "computer science?"

(b) In your opinion, what are the three main topics of computer science?

(c) In your opinion, what are the three main activities that people working in computer science carry out?

(d) What are the three "things" you like about computer science?

(e) What are the three "things" you dislike about computer science?

(f) In your opinion, what are the three main principles that should be watched in software development?

(g) Are you familiar with a disaster that occurred as a result of software failure? If yes, what was the disaster and what was the reason? If you are not familiar with a disaster that occurred as a result of software failure, can you think of any disasters that could occur as a result of software failure?

(h) In what follows, several concepts are presented. For each, please decide whether it is important ("why do we need it?" or "why do we not need it?") and explain why.
- Variable initialization
- Pointers
- Abstract data types
- Meaningful names (to functions, to parameters)
- Avoidance of code duplication
- Top down design
- Recursion

Part IV: Business Analysis of Software Engineering

Chapter 13, Software Project Estimation and Tracking
Chapter 14, Software as a Business
Chapter 15, The Internet and the Human Aspects of
　　　　　　Software Engineering

13 Software Project Estimation and Tracking

In This Chapter

- Introduction
- Objectives
- Study Questions
- Relevance for Software Engineering
- Poor Software Project Management
- Requirements
- Playing Games with Estimates and Deadlines
- Summary Question

INTRODUCTION

The Standish Group, in its much-quoted CHAOS report [Standish95], finds that software projects are often late and over budget. Project managers are frequently incapable of saving budget money; they spend it on late projects on extra personnel time. If time is indeed money, then the extra funding has already been used up on late projects. Managers without funds often adopt what we consider an inhumane practice, such as overtime, to compensate for lateness.

This chapter explores the effects of overtime on programmers. It then presents several more effective (we believe) methods of estimating and tracking time on task, so that managers who read this book will have additional tools to avoid being part of another negative statistic.

OBJECTIVES

- Readers will know what may be considered one of the most inhumane practices of software development.
- Readers will be able to use more accurate estimating and tracking tools.
- Readers will be able to play the Planning Game.

STUDY QUESTIONS

1. What are the main differences between COCOMO II and I?
2. How is overtime potentially harmful?
3. How does historical data help in estimating?
4. What is Clark's method of estimating?
5. How does the Planning Game work?
6. How are requirements derived from stories?
7. Why is part of the Team Software Process useful here?
8. How is Earned Value used in tracking?

RELEVANCE FOR SOFTWARE ENGINEERING

Every software project has a due date, even a surmised one. This chapter is about quick reactions to that date being perceived as late and several ways of handling these reactions. It is relevant to software engineering because producing software is labor intensive, and a book about the human aspects of software engineering should discuss scheduling in some depth.

POOR SOFTWARE PROJECT MANAGEMENT

It sometimes seems that managers estimate poorly on purpose. Programmers, thinking of their rank, try hard to meet impossible schedules. In this way managers can get more work from someone. It is frequently said that, "management is getting things done through other people." Under-estimation is a way of using every drop of creativity.

You can imagine the pointy-haired boss from Scott Adams' Dilbert™ cartoons descending the jetway from some trade show, whipping out his cellular phone, and calling one of his engineers. He informs the engineer that the company will build whatever impressive piece of software by the next trade show, thus setting an artificial due date for some arbitrary, dimly known piece of software.

It is almost certain that a few months later, someone will realize that the date cannot be met. Requirements could be scrubbed, but that means less functionality. Alternatively, some seemingly big items could be reduced in size, like testing, but that means less quality. In the absence of any good data, the boss will not budge from the earlier "estimation."

"Better, Faster, Cheaper"

A lot has been said in the industry recently about the "better, faster, cheaper" phenomenon, which is certainly relevant here. It once was that people would present these concepts by saying, "Better, faster, cheaper. Pick any two." Choosing to concentrate on two of these concepts made accomplishing the third difficult or impossible.

For example, producing something of high quality and within a deadline, is rarely inexpensive. The Apollo Lunar Landing program comes to mind. The lives of human beings were involved, so everything had to work, and work well. The date was seemingly arbitrarily chosen by the then-President, so no pushback was tolerated. It turned out to be quite expensive. Apollo was that rare government-sponsored program that had an indefinite budget. Therefore, it is an example of "better and faster," but certainly not "cheap" [Brooks79].

Software built quickly by small teams, regardless of the size of the product, rarely survives the rigors of constant use. Therefore, it is not "better," although it is "faster" and certainly "cheaper."

The final combination is better, and cheaper, but not faster. This is actually the combination that uses resources the most effectively. The "not faster" means that there is just enough time to do the work; therefore, adequate time is scheduled for the project.

Unfortunately, in these days of "Internet Time," the mantra of "better, faster, cheaper" means all three. Quality cannot be compromised, because customers abandon the source of poor quality; "faster" is the essence of the current era, as thousands of fast food franchises can attest. This leaves only "cheaper" as a factor for easy change, but few want to spend money for fear of not making sufficient profit. Hence, if the factor of speed can be increased, the goals will be met.

Perhaps in an attempt to deliver on these three aspects, some companies foundered. This might be one common reason for the rampant failure of dot-com companies.

TASK

Identify a product that is better and cheaper, but not faster. How does it reach its quality goals?

Overtime

Initially, the first reaction to a schedule shortfall was to add personnel. Fred Brooks pointed out the folly of this reaction course in [Brooks75], so this solution is probably wrong from a viewpoint other than cost. However, nowadays, increased cost is keeping some managers from remembering that. Many of them seem to consider software making as linear as ditch digging (see Chapter 1).

Therefore, the problem becomes centered on keeping costs down while increasing productivity. You can almost see the boss's hair twitch as overtime is expected until the delivery date. By using the same engineers for a little extra time, you save training costs, communications overhead, and extra benefits, since overtime often is not paid to salaried employees. Therefore, there is no perceived penalty to overtime.

This overtime may range from an hour a day to a half-day a week for over a year. Either way, it is chronic. One of the authors (Tomayko) observed a friend performing a half-day, long-term overtime in the nuclear power industry. Fortunately, reactor code written at 11:30 A.M. on a Saturday did not find its way into an American nuclear reactor!

Steve Maguire, once a Microsoft manager, observed those on chronic overtime as in the following story adapted from [Maguire94]: an engineer shows up for work at around 10:00 A.M., immediately processes electronic mail, and then works on his or her project for about a half an hour. Suddenly, the engineer realizes that his or her personal bills have not been done that month and then does bills until lunchtime. Feeling a little sluggish, the engineer runs with a friend at noon, showers, and eats a quick lunch. Going back to work about 2 P.M., the engineer first checks mail again, and then works for another hour or so. Getting sluggish once more, the engineer goes down the hall for some foosball or table tennis. By this time, it is 5 P.M., and the engineer realizes that not much time has been spent on the project. The engineer eats supper with his friends, then goes back to the office, and, with steely resolve, works on the project until midnight, then goes home and straight to bed. Nearly eight hours later, the engineer is up and at it again.

The engineer spends more than 14 hours on site, but only about half that time on the project. For the practitioner of chronic overtime, the things normally done in the "outside" world, like bills, exercise, and evening meals, become part of a normal day. It is no wonder Maguire starts his tenure on a new project by going down the office corridor at 7:00 P.M. to throw out the engineers and send them home [Maguire94].

It has been observed that an hour or two here and there, usually voluntary time on the part of an engineer, does increase productivity for a while [Beck00]. As might be expected, an extra half-day per week in an organization lacking flexible hours also causes a positive trend in productivity initially. However, chronic overtime in a flextime environment quickly causes engineers to revert to their original productivity, even though they are spending many hours at the job. Their brethren in the fixed-time company achieves the same result, just a little later.

What is wrong with this? Let us say that the engineer mentioned previously is married. If the spouse also works, most likely at specific hours, then the two may sleep in the same bed, but otherwise would not see each other until the weekend. If children are involved, it is highly likely they will not be seen as well. Those doing the extra half-day of work are absent from some family activities. This is clearly difficult for the engineers and their families.

Expecting this type of behavior as routine can cause a certain amount of resentment, erasing any positive effects of increased productivity. An author of this book (Tomayko) worked a six-week spurt of overtime once. His project delivered on time, largely through heroic effort on the part of the team. The software was never used. The fact that this is still bothersome after 20 years is a symptom of the scope of possible resentment.

TASK

List the pros and cons of daily overtime.

Avoiding Overtime

The best way to avoid overtime is to make better schedules and estimates of the time it is going to take to build the software. Time spent on planning is well spent if it is effective. The remainder of this section discusses several very effective ways of making estimates and tracking them.

Historical Data

Simple historical data is the basis of the first of these methods. We have all noticed that today's weather is much like yesterday's (in the same location). Building software is a lot like weather. Software in the same domain with the same functionality takes about the same amount of time to build. We observed a project manager moving a box containing prior estimates to a new cubicle. What had worked for two previous projects in the same domain now worked for a third. Eventually, the project manager transferred to a different domain. Certain software was new, but some had much in common with previous projects where the parts of the domain matched.

TASK

Identify two similar projects, and point out the parts of one that can serve as surrogates for parts of the other.

Clark's Method

In the case of many methods discussed here, they end their road to abstraction in a unit of software. For individual units, it is possible to be more accurate by using this equation:

$$LOC = (L + 4M + S) / 6$$

where M is how big (in lines of code) you expect the software to be, L is the biggest you can ever imagine it being, and S is the smallest you can ever imagine it being. For example: 102 (rounded) = (125 + 4 *100 + 85)/ 6 .

This relation is very close to the standard deviation. It was developed by an engineer named Clark for the Polaris submarine-launched ballistic missile project, one of the few government projects delivered ahead of time and under budget [Sapolsky72].

This equation and COCOMO I [Boehm81] were used among several hundred experienced practitioners. That use demonstrated the superiority of this relation, especially in the absence of experience with the product. In addition, this relation can be used to estimate the size of components, which estimates are then combined to make a more accurate overall estimate. This equation handles small units of code and scales up well.

TASK

Identify a software product in a domain that you know reasonably well. Divide the product into smaller parts. Use Clark's method to estimate the size of the parts. Add them together and compare the total size with the actual size. Comment on the accuracy of the estimate.

COCOMO II

Planning is difficult, and obviously error prone, in most methods. Many methods, even some of the most detailed [Humphrey95], start with the programmer making an estimate based on the gut feel of experience, and this is refined once the day's work is examined. The accuracy of the estimate is based on actual performance.

The shortness of a typical modern software engineering development cycle—two or three weeks—keeps the effects of mis-estimation small. Therefore, any refinement of the estimates must support cyclical development. It would be nice if the estimate had different levels of complexity. COCOMO II fills this bill [Boehm95].

COCOMO I is a component of many heavyweight processes. Even its original explication, [Boehm81], was huge. COCOMO I calculated project times and effort by the use of a volume metric, Delivered Source Lines of Code (SLOC). Thus arose an entire industry of size estimation as an input to COCOMO. It also caused endless debates over the definition of "lines of code." Do you count semicolons (logical statements) only? Include or exclude comments? On one hand, professors delighted in circulating an exercise to their classes resulting in a count of 1 to 10 lines, depending on what you consider as components of a line of code. At the other extreme is the Software Engineering Institute's guide to defining a line of code, more than a 100-page exercise [Park92]. Fortunately, the practice of having a coding standard simplifies defining a line of code.

COCOMO I was calibrated using only a few dozen projects in one company, TRW, and was highly inaccurate. Chris Kemerer did a study [Kemerer87] that showed that COCOMO estimates were about 600-percent low whether you were using Basic COCOMO (just the equations) or Intermediate COCOMO (raw estimates adjusted by applying a large number of qualifying factors). An interest group around COCOMO was formed and had annual meetings to swap data and recount struggles with COCOMO. Moreover, the technique is fairly useless for products under 10,000 lines of code.

After leaving TRW and a stint in government service, Barry Boehm, who originated COCOMO, accepted an academic appointment. By this time, the early 1990s, he noticed that the software world had changed, and that the waterfall model, which underlay COCOMO I, had been almost completely replaced by iterative models. These included his own, the Spiral Model [Boehm88]. Based on this observation, Boehm enlisted a larger number of companies to provide data on over twice as many projects as the COCOMO I. This data is used in calibrating a cyclic replacement to his original model, called COCOMO II [adapted from Boehm95].

Boehm realized that the ease of the use of late-generation languages and other tools would make three layers of practice necessary. At one extreme is end-user programming, made possible by scripts used in programming spreadsheets, query systems, and planning systems. The other end is infrastructure, like operating systems, database managers, and networking systems. In the middle are application generators and system integrators. The estimation needs of each of these groups are different, so there are different layers to COCOMO II.

Therefore, there are three stages to making a COCOMO II estimate: Application Composition, Early Design, and Post Architecture. The Application Composition model also supports prototyping at any point in the life cycle.

The end-user programmers are unlikely to need anything as detailed as lines of code, both for the simple reason that they just do not need the fine-graininess and because their development cycles are so short that they don't have the time to develop a detailed estimate. Prototyping is the same. Therefore, the Application Composition model uses Object Points, not to be confused with object-oriented development "objects." They may turn out to be roughly contiguous with such objects, but they aren't necessarily the same.

There is a seven-step process in coming up with this estimate. First, count the estimated numbers of screens, reports, and third-generation language objects. Second, figure complexity using Table 13.1.

TABLE 13.1 Determined Complexity According to the Estimated Numbers of Screens and Reports

	For Screens				For Reports		
Number of Views	Number and source of data tables			Number of sections	Number and source of data tables		
	Total<4 (<2 servers <3 clients)	Total<8 (2/3 servers 3–5 clients)	Total 8+ (>3 servers >5 clients)		Total<4 (<2 servers <3 clients)	Total<8 (2/3 servers 3–5 clients)	Total 8+ (>3 servers >5 clients)
<3	Simple	Simple	Medium	0 or 1	Simple	Simple	Medium
3–7	Simple	Medium	Difficult	2 or 3	Simple	Medium	Difficult
>8	Medium	Difficult	Difficult	4+	Medium	Difficult	Difficult

The third step is to attach weights to each number in the cells (Table 13.2).

TABLE 13.2 Complexity-Weight for Screens, Reports, and Third-Generation Language Components

Object Type	Complexity-Weight		
	Simple	Medium	Difficult
Screen	1	2	3
Report	2	5	8
Third Generation Language component			10

The fourth step obtains the Object Point (OP) count by adding all the weighted object instances. Then, estimate the percentage of reuse. Use the percentage of reuse in this equation to get the New Object Points: NOP = (OP − % reuse)/100. The sixth step is determining productivity from Table 13.3. This corresponds to *velocity* in eXtreme Programming (XP).

TABLE 13.3 Productivity According to Developer Experience and Ability and Development Environment Capability and Maturity

	Very low	Low	Nominal	High	Very high
Developer experience and ability	Very low	Low	Nominal	High	Very high
Development environment capability and maturity	Very low	Low	Nominal	High	Very high
PRODUCTIVITY	4	7	13	25	50

Finally, based on PRODUCTIVITY = NOP/Person-Month, compute the person-months: Person-Month = NOP X PRODUCTIVITY.

The Application Composition model provides enough structure for the estimations necessary in the XP Planning Game. The Early Design model uses function points, and the Post Architecture model uses COCOMO I. The frequency of XP cycles makes the use of these additional components of COCOMO II too difficult. However, those interested in function points are directed to [Albrecht and Gaffney83]. Those who want to use COCOMO I, see [Boehm81].

Basically, in a noniterative production schedule, Application Composition estimates are done early in the software development life cycle, function points after higher-level analysis, and COCOMO I after detailed design. Therefore, there are at least three iterations to the estimate.

TASK

Take a software product and try to identify the number of person-months to build it using COCOMO II.

Earned Value

So far, we reviewed some estimation methods. Before moving on, we discuss what some consider a very accurate tracking mechanism. Tracking software development is important. Estimates that are clearly wrong have to be redone. At all times, consistency with the budget must be tracked.

Currently, this tracking is done with status meetings. Some projects use weekly status meetings and some wait for several weeks to elapse. Obviously, frequent iterations tend to favor weekly meetings.

In Earned Value estimates, an engineer does not get credit for something until it is completely finished. No more "90 percent done." And 100 percent of a component may only be a small percentage of a product.

Let us consider this in more detail. Think of a product with 20 roughly equal parts. If we consider percentages again, nine components completed 100 percent are 45 percent of the product. However, we might actually be farther along than 45 percent, as we may have done some work on other components. When all components are 100 percent done, the product is 100 percent done. If you are going to use this tracking method, it is important to explain it to the clients, especially ones who are used to hearing developers say that they are "90 percent done" for half the length of the project. Otherwise, they will not understand why you've been working for months and are a small percentage finished.

Let us take an example from our 20-item product. If there is an overall architecture, and this is a deliverable, the product probably *is* 90 percent done, or we would not know what to do next. It is not completely done, however, so we get no credit for it, making some obvious work seem untracked. It is clear that we should use abstraction more extensively, to make the components stand alone. Therefore, in the waterfall life cycle with some feedback, we would be 0 percent done until the feedback ceases. That would be very strange to most clients!

Iterative methods are favored when using earned value to track, because you tend to be "finished" in a short time because small components are developed. This forces the client to divide the problem. The role of "tracker" in agile processes is easier using earned value.

TASK

Pretend you are walking down the hall to a project status meeting with the client. You have been using earned value to track progress in the project. Explain how it works before you reach the meeting room.

The Planning Game

The Planning Game of XP has an entire book devoted to it [Fowler01]. Here we will follow the inputs and outputs to the Planning Game in some detail. Hopefully, in this way, it can become obvious how to use it for better estimation.

Initially, everyone—the clients, the developers, and management—agrees on the length of the iteration, usually two to six weeks. This roughly sets the amount of tasks that can be done.

The client prepares story cards. These can be both functional and nonfunctional. For example, "The user selects what function to execute," or "the functions are presented in a clear manner." Requirements can later be derived from these stories. For example, "selects" indicates that there is more than one function, and "clear" is a nonfunctional quality attribute, presently undefined in detail.

The client prepares enough of these story cards to cause slightly too much work for this iteration. When these cards are initially handed to the developers, they are the first "move" in the Planning Game. The second "move" is considering velocity (see Chapter 2, "Software Engineering Methods")—the developers estimate the time needed to do each story. Since each developer has a slightly different velocity, the person estimating the work takes ownership of the task. Developers can use either Clark's Method or COCOMO II to estimate the size or time it would take to do the task implied by the story.

These results are then returned to the client. The client considers the market and the length of the tasks and returns the set of stories to the developers in a prioritized manner.

The developers spread out the prioritized tasks. Some are done immediately and some are delayed indefinitely. For example, the "clear" story may wait for the look and feel of the product to be decided in a later iteration when all functions are done, or the look and feel may be done immediately.

The developers split the stories up by ownership and iteration time. Some tasks might change hands at this point. Note that the tasks may not equal the stories. For example, consider a story's function. It may take several abstractions to achieve that function; therefore, several tasks. The tasks are noted on the story cards. It could result that the same programmer does all the tasks, or some are split off at this point.

We want each developer to finish at about the same time. This time is the length of the iteration. The customer is asked if this iteration is too long to return business value. If it is, then some tasks are delayed. The idea is that the amount of time for iteration and to complete most of the stories should match and be short in duration.

The developers then work on the code, implementing functions and quality attributes defined by the stories. When they and the iteration are finished, the product is delivered to the customer for release. The income from this release can partially fund later releases.

TASK

Conduct the Planning Game for two iterations of developing software for an automatic teller machine (ATM).

REQUIREMENTS

As has been mentioned in Chapter 4, "Software as a Product," requirement formulation has a significant influence on software success. In this subsection, we highlight several issues related to requirements elicitation.

The Method of Up-Front Requirements Elicitation

Current requirements elicitation methods reinforce improper beliefs. Many projects try to follow the Waterfall Software Life Cycle and other, usually linear, software development, life-cycle models. The requirements are gathered first in one big effort [Davis90]. Frequently, these requirements turn out to be rife with omissions and misconceptions. Correcting them costs time and money. The result has been a movement toward more iterative models [Tomayko00]. The history of iterative requirements gathering has itself gone through several cycles, of which agile methods are the latest (See Chapter 8, "History of Software Engineering"). The trouble is that with each new iterative method, the requirements elicitation process appears to become less settled.

The attitude toward requirements gathered early in the process is incorrect in that ones missed at that stage are considered defects when added later. The entire point of iterative processes is that requirements are seldom omitted; they are just unknown. There is simply no way that all the requirements of even well-understood problems could be known. Therefore, why even try? Requirements are elicited by agile methods in a more practical way, and thus estimations are better.

Requirements Elicitation in Agile Methods

It seems difficult to identify requirements from stories. However, research has shown that an "agile attitude" toward requirements is a very effective means of acquiring them [Smith01], and thus doing repeated accurate estimations.

In agile methods, the "user stories" are just the beginning, points of both the requirements gathering and development processes, and thus of estimates. Early requirements are simply a place to start. One is expected to add more requirements as more is known about the product. Conversely, the method of trying to gather all the requirements before starting development will almost certainly result in errors and surely takes too long. In such a development process, the client is prompted by the product to "remember" some things and by the marketplace to want others changed. "I'll know it when I see it (IKIWISI)" [Boehm00] has become a well-known requirements identification method. In effect, the early version of the product becomes a prototype. Agile methods are designed to appeal to clients that insist on IKIWISI. Moreover, in this way, the client is kept from the responsibility of "getting the requirements 'right'." There are no wrong requirements; there are simply some waiting to be discovered.

Difficulties for Estimation Caused by Agile Methods of Gathering Requirements

This agilist attitude toward requirements makes estimation and software architecture development more difficult, and verification easier, than traditional methods. Without knowing the final form of the product, or marketplace demands, large-scale product estimation, like COCOMO I, is going to be impossible to get correct [Thayer97]. It is little comfort that requirements omissions and changes caused by reacting to the competition make most estimates incorrect right now. Agile methods are likely to be right about the costs involved in the current cycle, but estimating is poorly understood for the unknown requirements of the next cycle.

As mentioned earlier, one thing that can be done is to fund the project one cycle at a time, which is equivalent to funding an entire project using an older development method now. There will be times, however, when knowing the total cost is necessary, as in contract work. In these cases, the customer expresses the requirements as well as it can, and the estimate is adjusted by the probable cost of later changes. For example, if a project is estimated at $1 million, and prior projects of roughly those same characteristics have had the cost of "changed" requirements at around 20 percent, then the estimate is $1.2 million. Of course, as with all estimations, this method cannot be used without considerable historical data.

As for the architecture, that chosen by the team during the early cycles may become just plain wrong, as later requirements become known. Rework of the architecture matches the refactoring principle. One developer identified refactoring as rework, with its attendant negative properties, such as deleting some code, notably increased cost. Either way, significant refactoring is to be expected in an atmosphere where requirements are relatively unknown. Confidence in the requirements translates to confidence in the architecture.

TASK

List five client stories from a piece of software. Derive requirements from them.

The Team Software Process Development Manager

A software development process usually associated with heavier weight methods is the Team Software Process (TSP). However, the role of Development Manager as found in TSP is useful in the estimating and tracking methods just described. The primary purpose of the Development Manager is to decide the overall strategy and what will be done in each iteration [Humphrey00].

In other words, TSP depends on cyclical development, much like the more modern development methods we described previously. Perhaps, with the team coach or a lead developer, the client can take on the role of Development Manager.

The basis of the role is as follows: the Development Manager obtains the goals for the entire product and figures out how many iterations are needed to develop the product and what their content is. This is tantamount to obtaining all the stories known at the beginning, and then choosing the subset needed for the first iteration.

TASK

Develop a strategy for making software for an ATM.

PLAYING GAMES WITH ESTIMATES AND DEADLINES

Probably the clearest example is creating slack time for a project. The worst case of poor estimating is when the manager doing the final estimate does not allow any time for false starts or even bathroom breaks. This is almost as bad as missing Christmas because you did not get the memo indicating the date!

Slack has a negative connotation, indicating not paying attention to something or laziness. This is unfortunate, as it is the lack of slack time that causes many schedule, and thus cost, overruns [DeMarco02]. The XP concept of "velocity" is not a part of these estimates. Therefore, a meeting called suddenly by the boss causes the product to be as late as the meeting is long.

How do conscientious managers keep slack in their schedules? By keeping two sets of books; essentially, lying. As an example, let us take a short development cycle of three weeks. Let us say that there is enough slack time in the tasks and that the cycle, with slack, is four weeks long. The manager knows that a schedule can be a self-fulfilling prophecy [Abdel-Hamid89]. Therefore, the manager tells the developers three weeks and upper management four. This way, the manager can be budgeted for the four weeks the bosses think it will take, while the development team will strive to be done in three weeks. When they eventually finish, in three weeks and a couple days, the developers are not too surprised by a little lateness in schedule, and upper management is very happy that they saved money. The two-faced manager looks like a scheduling genius.

Eventually, upper management notices this team never uses all the funds allotted to it, and the developers realize that they can be late a couple of days without serious repercussions. Pressure is exerted in both directions. The manager in the middle appears to be like other colleagues in poor estimation again. The game is over.

How much slack is needed? The difference between velocity and full-time work. The developers can never be 100 percent time on task, or they would not be human. For example, one of the authors (Tomayko) once worked for a large software development organization. He noticed some workers who took forever, it seemed, to get started in the morning. These persons would talk to those already working by peeking over cubicle walls, coffee cup in hand. Of course, little or nothing would get done during that period. There were also breaks and lunch. Velocity is a concept that resulted from managers and programmers realizing that a day is not a full day.

There is some slack time needed even if velocity is used. As an example, some time must be scheduled for rework if software inspections are to be used. There is no way of knowing how long rework will take, since the results of inspections are not known when they are scheduled. One solution is to look at the historical data on the length of such work. The game described previously can be played for this effort. Other slack time is needed for change. In an XP project, the cycles are kept short and unit tests are part of development, thus minimizing the effects of change.

Again we ask, how much slack? And, again, we can use historical data. Of course, we assume that a CIO does not get off a plane from a conference and say, "We have to have X by this time next year."

TASK

Make a one-month schedule for a software development project with appropriate slack time for each task.

SUMMARY QUESTION

Develop a long- and short-term estimation plan for a project, being explicit about how you would do estimations.

FOR FURTHER REVIEW

Find a scheduling disaster in your company. Analyze it. What went wrong?

REFERENCES

[Abdel-Hamid89] Abdel-Hamid, Tarek K., "The Dynamics of Software Project Staffing: A System Dynamics Based Simulation Approach." *IEEE Transactions on Software Engineering* 15(2) (1989): pp. 109–119.

[Albrecht and Gaffney83] Albrecht, A. J., Albrecht and Gaffney, J., "Software Function, Source Lines of Code and Development Effort Prediction," IEEE Transactions on Software Engineering, SE-9, Volume 6, (1983): pp. 639–648.

[Beck00] Beck, Kent, *Extreme Programming Explained*, Addison-Wesley, 2000.

[Boehm81] *Software Engineering Economics*, Prentice Hall, 1981.

[Boehm88] Boehm, Barry, "A Spiral Model of Software Development and Enhancement." *Computer*, Volume 21, Issue 5, (May 1988): pp. 61–72.

[Boehm95] Boehm, Barry; Clark, Bradford; Horowitz, Ellis; and Westland, Chris, *Cost Models for Future Software Life Cycle Processes: COCOMO 2.0*, USC Center for Software Engineering, 1995.

[Boehm00] Boehm, Barry, "Requirements that Handle IKIWISI, COTS, and Rapid Change," *Computer*, IEEE, (July 2000): pp. 99–102.

[Brooks75] Brooks, Fred, *The Mythical Man-Month*, Boston, 1975.

[Brooks79] Brooks, Courtney G.; Grimwood, James M.; and Swenson, Loyd S. Jr., *Chariots for Apollo: A History of Manned Lunar Spacecraft*, NASA SP-4205 National Aeronautics and Space Administration Scientific and Technical Information Office, 1979.

[Davis90] Davis, Alan, *Software Requirements*, Prentice Hall, 1990.

[DeMarco02] DeMarco, Thomas, *Slack*, Broadway Books, 2002.

[Fowler01] Fowler, Martin, and Beck, Kent, *Planning Extreme Programming*, Boston, MA, 2001.

[Humphrey95] Humphrey, Watts, *A Discipline for Software Engineering*, Boston, MA, 1995.

[Humphrey00] Humphrey, Watts, *Introduction to the Team Software Process*, Boston, MA, 2000.

[Kemerer87] Kemerer, C. F., "An Empirical Validation of Software Cost Estimation Models," *Communications of the ACM* 30 (5), (May 1987): pp. 416–429.

[Maguire94] Maguire, Steve, *Debugging the Development Process: Practical Strategies for Staying Focused, Hitting Ship Dates, and Building Solid Teams,* Microsoft Press, 1994.

[Park92] Park, Robert, *Software Size Measurement: A Framework for Counting Source Statements*, CMU/SEI-92-TR-020, 1992.

[Sapolsky72] Sapolsky, Harvey M., *The Polaris System Development; Bureaucratic and Programmatic Success in Government,* Harvard University Press, 1972.

[Smith01] Smith, John, *A Comparison of RUP and XP*, Rational Software White Paper, 2001.

[Standish95] *www.scs.carleton.ca/~beau/PM/Standish-Report.html.*

[Thayer97] Thayer, Richard, and Dorfman, Merlin, eds, *Software Requirements Engineering*, IEEE Computer Society Press, 1997.

[Tomayko00] Tomayko, James, "An Historian's View of Software Engineering," *Procedings of the IEEE Conference on Software Engineering Education and Training,* 2000.

14 Software as a Business

In This Chapter

- Introduction
- Objectives
- Study Questions
- Relevance to Software Engineering
- A Brief History of the Software Business
- Summary Questions

INTRODUCTION

This chapter is about software as a business. It consists of two main parts: a brief account of how software became profitable and recent stories of making money with software.

OBJECTIVES

- Readers will become familiar with the main concepts related to the business aspects of software.
- Readers will be aware of basic operations that should be carried out when one considers the development of a new software product.

STUDY QUESTIONS

1. Why was software not marketable at first?
2. How did one make money with software at first (and even later)?
3. What is a "stock option?"
4. What is the most ubiquitous form software takes today?
5. How do humans affect the current form of software?

RELEVANCE TO SOFTWARE ENGINEERING

The *raison d'être* for software is to earn money. Profit was not always possible, since hardware concerns dominated the business. Even now, several years after what is called in America "the dot-com bust," some of the most famous people in computing are so well known because of software they help produce.

A BRIEF HISTORY OF THE SOFTWARE BUSINESS

In the beginning, hardware concerns far outweighed software worries. In Chapter 8, "The History of Software Engineering," we saw that the EDSAC came to Cambridge essentially without software. Researchers wrote their own programs for the machines. Most users of computers in the 1950s also wrote their own software. This method quickly became expensive even for the big U.S. government contractors. User groups sprang up, like the aptly named SHARE, to exchange programs and ideas. SHARE was primarily made up of IBM (International Business Machines) users. The members gave each other software and asked questions of each other and the manufacturer.

There was only one way to make money in software during the 1950s and 1960s: as a contractor. Some contractors were independent, but they could not do the really big projects. Some groups came from a manufacturer's programming team. A good example is the joint IBM-American Airlines group that built the SABRE airline reservations system in the early 1960s. The IBM people knew the hardware; American employees knew what the problem was. Some companies were large system contractors independent of manufacturers. Another example is the Systems Development Corporation (SDC), which wrote most of the software for the huge SAGE air defense system.

Smaller, but very capable, specialty shops contracted for certain jobs. For example, the Charles Stark Draper (then "Instrument") Laboratory of MIT did the software for the guidance system for the Polaris missile, the Apollo spacecraft, the F-8 *Crusader* fly-by-wire aircraft, and the Space Shuttle backup computer. To this day, nearly 50 years later, it still does U.S. government contracts. Over the years, Draper developed a style (cyclic, rather than asynchronous priority driven) for operating systems. It also was able to reuse designs and code among its many projects.

The 1960s were a wake-up call to the computer industry. Some software projects began to be late and dysfunctional. A classic example is the IBM System 360 operating system, then one of the biggest software projects ever. Project leader Fred Brooks wrote his popular book *The Mythical Man-Month* [Brooks75] using material from this effort. Draper's Apollo software was also considered a mess, obscured by the 1967 fire that took the lives of three astronauts, and the subsequent nearly two-year redesign of the Apollo Command Module. At the same time, across the MIT campus, research work was underway on the MULTICS time-sharing operating system.

These groups of users, independent contractors, and manufacturing company software groups attended the first software engineering conference in Garmisch, Germany (see Chapter 8). That conference, coupled with some other events, changed the nature of software as a product forever. IBM 360s were very common, the Apollo Program culminated in a lunar landing, remote entry and time-sharing systems were stable, and Moore's Law for hardware made things easier on software. Moore's Law says that the power of hardware, both memory and speed, doubles every 18 months. Software got bigger to take advantage of this trend.

Another event of the late 1960s that had a huge impact on the software business was the decision of IBM to "unbundle" software from their hardware and sell it separately. Since this was the age of IBM, all its competitors changed their policies toward software in the few months after IBM's unbundling decision on June 23, 1969 (to take effect January 1, 1970). IBM also reduced hardware prices by 3 percent.

The seemingly paltry 3 percent reduction in hardware costs seemed to belie the reason for the unbundling: the cost of making software. This is actually a correct percentage. Even rolling in such costs as software research and development, few computers had software costs more than 5 percent of the total. There were no manufacturing or marketing costs [Campbell-Kelly03].

Some companies entered the market to sell software for large computers. Things such as sorts, searches, and databases had a lot of competition. Some software from that era, like IBM's CICS, is still in use today.

Microcomputers began to be sold in the 1970s. As the next decade began, big computer manufacturers like IBM and NCR began their own microcomputer projects. The industry, expecting IBM to sell the most personal computers, awaited its choice of operating systems.

At the time, the CP/M microcomputer operating system, built by Digital Research, was considered to be doing well in the marketplace. IBM chose DOS (Disk Operating System) by the then little-known Microsoft. IBM-style personal computers (often called PCs) took over the market as a standard. Soon, IBM opened its architecture to the other companies. The PC took command and Microsoft suddenly sold, it seemed, to everyone. Apple Computer's famous 1984 Super Bowl advertisement caused a brief flurry of activity around its Macintosh and its proprietary operating system. It never acquired more than a small part of the market. Even today, although the G5 64-bit dual microprocessor debuted as the fastest machine of its type, it is difficult to find software for it. Microsoft made versions of its Office suite of programs (Word, naturally, a word processor; PowerPoint®, a viewgraph tool; Excel, a spreadsheet; and others) available for the Macintosh quite soon after it came out.

Unlike large computers, microcomputer sales took off in step with software. The VisiCalc spreadsheet program was the first popular software. It was available on early PCs. The sales of millions of computers are now possible due to the availability of free software network browsers. The World Wide Web as a whole is the "killer app" of today. Companies like Gateway and Dell sell microcomputers on the basis of easy Web access and the Microsoft® Office suite of programs.

The rapid expansion of software creation in the 1990s meant that computer programmers were in short supply. This became part of the "dot-com" boom of Internet companies. One compensation method used by new companies was stock options. Employees would have periodic opportunities to invest in a growing company by purchasing its stock at deeply discounted prices. Thus, even secretaries could, and did, become millionaires [Ceruzzi01].

However, the general economic downturn and the growth of outsourcing (see Chapter 6, "International Perspective on Software Engineering") kept the software market from totally rebounding. Another big driver of the industry became, ironically, open-source, or free software. This hearkens back to pre-1970 IBM. Even more ironically, IBM is one of the biggest distributors of one of the most successful Open Source products, Linux, a modification of the Unix operating system begun by Linus Torvalds.

There are many Open Source products. Most have a group of software specialists who are the only ones able to touch the source code. This committee of programmers is bombarded with suggestions for improvements and defect fixes. Open Source products change regularly.

Software product producers and consultants (even if they only advise on how to use free products!) can still make money in software. We discuss how in the next section.

TASKS

1. Explain how software became a product.
2. What are the two ways of making money in software?
3. Explain how stock options work.

Time-to-Market

Perhaps the most important trait of software products nowadays is the time spent in development. At one point, it was "better, faster, cheaper; pick any two." Now, software products must be *all* three. This means that there has to be a high-quality product made inexpensively and quickly. No more (to quote the commercial) "we will sell no wine before its time." There is a concept known as "Internet Time," which means hardly any time from development to market for software.

There have been software development life cycle changes to capitalize on this concept. "Rapid Application Development" is one fashion [McConnell96]. However, one of the easiest developments to use for this purpose is eXtreme Programming (XP). Its explicit attention to returning value to the customer with each release makes it a perfect tool for speed.

Business value seems to have been finally discovered by the software industry. Organizations like the Software Engineering Institute (SEI) are paying attention to business value since many of their software architecture products have a business basis.

Basically, time-to-market is related to market share. This is truer now than in the "nobody got fired for buying IBM" days. No company is exempt anymore. Products delivered according to standards, rather than their speed of development, drives many customers.

The concept of market share means the early capture of a percentage of the market by an early release of software. Users become enamored with the functions in this version and willingly buy later versions. Thus, functions and qualities of initial releases must be good. Surrendering quality at this point is self-destructive. If the company is perceived as skimping on quality, it will not make further sales.

TASKS

1. What does "market share" mean?
2. Why is it important to maintain quality while gaining market share?

Business Cases

Nothing is sadder (or seemingly more prevalent recently) than the "one great idea" companies. You know them. They are the ones begun by some brilliant engineer with an innovation that solves a problem that does not exist. Many bankrupt companies have this orientation. Most believe in the adage "build a better mousetrap, and the world will beat a path to your door."

This is not always true. Perhaps a better restatement is, "if the world wants a mousetrap, and you have the best one, the world will beat a path to your door." The best designs solve a problem people did not know they had. The worst solve a problem that does not exist.

The alternative to entering the marketplace blindly is making a business case for a product. The problem the product is trying to solve is clearly identified. The ways in which the product solves the problem is documented. A business case can serve as the basis for the requirements. Both functional requirements and quality attributes are discovered.

When a new product is envisioned might be a good time for a "focus group" of probable users (a small group of less than 10 people) to tell developers what they are looking for. This meeting can serve as the source of user stories, prioritization of requirements, and the need for new requirements.

TASKS

1. What are business cases?
2. What are focus groups?
3. How are requirements ranked by priority?

Business Plans

Once the business case is made, a plan is developed for the company that will make the software. Let us assume this plan will contain a vision for the product(s), site of the company, a review of the competition, and marketing plan.

Vision: This is a section of the document that tells what the product will look like and what it will do. In other words, what problem it is likely to solve. For example, a vision might be increased fuel mileage. By generalizing the vision thus, the objective can be achieved either by automating the fuel/air mixture or by some other way of saving gas. We do not concern ourselves with "how" at this point, just "what."

Site: A department store might need a corner lot downtown or prime space in a busy mall to generate enough traffic to break even or make a profit. Sites are just as important to the software business, but for other reasons. Alternatively, a more established company may have several sites already, increasing the value of remodeling, since workers want to come to a pleasant place to spend their day.

Review of the competition: This is the most important part of the business plan in general. It is assumed that the chief characteristics of your product have been determined. These characteristics are then compared to those already in the marketplace or definitely announced. In the 1960s, IBM was infamous for announcing that certain computer functions would become available long before they did, thus freezing the "no one ever got fired for buying IBM" crowd. Meanwhile, other companies rushed their products to market. The CDC 6600, thought by some to be a superior computer to the IBM 360, died in the marketplace while CIOs waited for IBM. The CDC 6600 was the basis of a lawsuit against the IBM 360, model 91 [Ceruzzi00].

Each and every threat to market share is listed and analyzed here. You need to know, in detail, who the enemy is and what their chief weapons are. Then you can plan to counter with your product. Finding nothing here is akin to not having a problem.

Marketing plan: Many computer practitioners who think of a good idea forget the key activity of marketing. Even if you have a "better mousetrap" and a mouse problem, if you don't bring them together, nothing results. Sales are very important; they are the lifeblood of a business. Many believe that Thomas Watson Sr. realized this, emphasized it, and made IBM into a market leader.

These are the parts of a decent business plan. Several concepts may seem obvious, but they are put here as reminders. Some are so obvious that they are frequently overlooked.

TASKS

1. What are the key components of a business plan?
2. Why is a marketing plan important?

Statement of Work

If a product is going to be made for a single procurer (contracting), then a Statement of Work (SOW) may take the place of parts of the Business Plan for that software product. One of the most important things included here is the "customer liaison," a person who will handle communications with the client and will be familiar with many ideas in this book. They are the human face of the software project.

The customer's response time ought to be fixed in a SOW. In other words, the time given to the client to think of the answer to a question should be determined and be constant. This is very important in "out of sight" development to keep it from becoming "out of mind" as well. It is amazing that many companies will invest a lot of money in contract software, but then ignore requests for requirements clarification. This is probably one big reason why many custom software projects wind up considered failures. It is certainly why XP advocates (see Chapter 2, "Software Engineering Methods") want to keep delivering business value through having an on-site customer.

TASKS

1. What is a Statement of Work (SOW)?
2. What is the most important part of a SOW?

SUMMARY QUESTIONS

1. How did software become a commodity?
2. How can you always make money in the software business?

FOR FURTHER REVIEW

Prepare a business plan for an actual or a fictional software product.

REFERENCES AND ADDITIONAL RESOURCES

[Brooks75] Brooks, Frederick, *The Mythical Man-Month,* Addison-Wesley, 1975.
[Ceruzzi00] Cerruzi, Paul, *A History of Modern Computing,* MIT Press, 2000.
[Ceruzzi01] Cerruzi Paul, personal communication, 2001.
[Campbell-Kelly03] Campbell-Kelly, Martin, *From Airline Reservations to Sonic the Hedgehog: A History of the Software Industry,* MIT Press, 2003.
[McConnell96] McConnell, Steve, *Rapid Development: Taming Wild Software Schedules,* Microsoft Press, 1996.

Head, Robert V., "Getting SABRE off the Ground," *IEEE Annals of the History of Computing* (2002): pp. 32–39.
Kemerer, Chris F., *Software Project Management: Readings and Cases,* Irwin, 1997.
Kotler, Philip, and Armstrong, Gary, *Principles of Marketing,* Prentice Hall, 2002.

15 The Internet and the Human Aspects of Software Engineering

In This Chapter

- Introduction
- Objectives
- Study Questions
- Relevance for Software Engineering
- E-Commerce
- The Timeless Nature of the Internet
- Summary Questions

INTRODUCTION

This chapter examines several Internet-related topics of which each software engineer should be aware. The chapter consists of two parts.

The first part introduces the main concepts, attitudes, and principles of e-commerce (electronic commerce). Although this section does not address the human aspect of software development directly, we include this topic in the book since many of currently developed software systems are constructed to support online businesses. Consequently, software engineers should be aware of the different aspects of this topic.

E-commerce is a broad area that consists of and is connected to many things. For example, connections can be found between e-commerce and topics such as marketing, economy, human-computer interface, management, and of course, the development of Web sites that support e-commerce transactions. Furthermore, e-commerce can be discussed from different perspectives, such as financial implications, models of e-commerce, and possible online business activities. Naturally, this chapter does not examine all these topics and perspectives. In line with the spirit of this book, this chapter encompasses some of the human aspects of e-commerce. Specifically, it analyzes from a cognitive and social perspective topics related to the end users, the ones who use the developed applications for online purchases. We also mention relevant topics related to the developers and the management of e-commerce companies.

The second part of this chapter focuses on the timeless nature of the Internet and its application to the human aspects of software development. Specifically, we describe several events in which this nature of the Internet may be useful.

OBJECTIVES

- Readers will become familiar with the main concepts and principles related to e-commerce and their connections to the human aspects of software engineering.
- Readers will become familiar with examples of e-commerce companies, learn principles for analyzing such companies, and gain basic ideas of what topics to consider when and if they decide to develop a Web site that supports an online company.
- Readers will be able to examine e-commerce environments from cognitive and social perspectives.
- Readers will be able to identify the advantages and disadvantages of the timeless nature of the Internet.

STUDY QUESTIONS

1. What is e-commerce?
2. Have you ever purchased a product online? If yes, list motives that encouraged you to buy online. Did the purchase go smoothly? Did you face any difficulty?
3. Study several e-commerce Web sites (Amazon, for example).
   a. Describe the Web sites.
   b. Find data about the companies. Compare this data with data about similar businesses that

           i. Do not offer online purchasing.
           ii. Combine e-commerce with the traditional way of doing business.
 4. Identify main features of e-commerce Web sites.
 5. How do e-commerce companies present their merchandise? Is this way different from the way traditional companies present their merchandise?
 6. What advantages and disadvantages does e-commerce have over traditional commerce?
 7. Suggest possible connections between e-commerce and the Code of Ethics of Software Engineering.
 8. Suggest possible connections between e-commerce and education in general and the education of software engineers in particular.
 9. There are different e-commerce business models. This question focuses on the B2B and B2C (Business to Business and Business to Customer, respectively) models.
      a. For each of these two models, find two examples of online companies. Describe differences and similarities between the Web sites of these companies.
      b. Some argue that e-commerce fits the B2B model especially well because of the relative ease that it offers in comparing huge numbers of prices. In your opinion, does this argument appeal also to private customers? How can you validate your opinion?
 10. If you wanted to establish an e-commerce company, what would be your first steps?
 11. Why and how can software developers use the timeless nature of the Internet for their everyday work?
 12. Find connections between the timeless nature of the Internet and the content of Chapter 3, "Working in Teams."

RELEVANCE FOR SOFTWARE ENGINEERING

In this chapter, we focus on two aspects of the Internet. The first relates to the end users—the people who use the e-commerce online applications. In this sense, we close the circle of developers–clients–end users. Accordingly, while the interrelation between developers and clients was discussed in Chapter 4, "Software as a Product," in the first section of this chapter we explore possible connections between developers and end users by addressing cognitive and social aspects of e-commerce.

As is expected from such a rich topic as e-commerce, an interdisciplinary approach is taken in this chapter. Since it is impossible to review in this book all the relevant issues associated with e-commerce, we leave for the readers the learning of parts of the topics, by posing questions that may direct them how to further explore those directions they find interesting.

The second aspect looks at the Internet through one of its salient attributes—the timeless nature of the Internet. It explores how different processes have been changed during the past decade in which the Internet entered our life. Clearly, this nature of the Internet is of high relevance for software engineers. First, it may influence their everyday work; second, it may influence the type of applications they are developing.

E-COMMERCE

This section starts by explaining what e-commerce is. Then it examines e-commerce from cognitive and social perspectives.

What Is E-Commerce?

E-commerce is the arena in which products are sold directly over the Web. From the human perspective, people who do e-commerce conduct business by using online channels. One immediate outcome of this feature of e-commerce is that the buyer and the seller do not see each other when the transaction is conducted. Furthermore, the seller and buyer exchange neither cash nor credit cards. All the purchase activities are performed electronically, except, of course, the shipping of the merchandise itself (when it cannot be delivered online, unlike many software products).

E-commerce started around 1995, almost in parallel with the exposure of the Internet to the wide population. As soon as the commercial potential of the Internet was recognized, new e-business businesses were established and traditional businesses added the electronic selling channel. The dot-com era (in the 1990s) was a record in this movement (see Chapter 6, "International Perspective on Software Engineering").

Although there are differences between traditional purchase style and online purchasing (in online shops one cannot touch the products one wants to buy), even a surface examination of the features of the Internet reveals its commercial potential. For example, businesses can reach potential customers even if the customers do not live nearby; customers can reach shops that are not located near their homes; all online shops are open all the time; many online shops enable their customers to read reviews by other customers. It seems to be a win-win situation from both the customer's and the businesses' perspectives. The expectations of this selling channel have not yet been fulfilled, however, and many customers still prefer to shop in real shops, to touch what they intend to buy and not to skip the shopping experience. Still, there are businesses for which the online channel is their main business model.

Table 15.1 adds to the preceding discussion by mapping seller-buyer relations according to their synchronisms and asynchronisms with respect to place and time. For example, in traditional shops, both the place and the time are synchronous. In e-commerce, both the place and the time are asynchronous.

TABLE 15.1 Transaction Styles: Time and Place Synchronism

	Time	
Place	**Synchronous**	**Asynchronous**
Synchronous	Shop	Lottery
Asynchronous	Purchase by the telephone	E-commerce

TASKS

1. In your opinion, what kinds of businesses and goods are fit for e-commerce? Why? Consider in your answer also products that you would not naturally purchase online (food, for example).
2. Should the Web sites that support e-commerce businesses be different for each type of product? Explain your opinion.
3. In your opinion, what types of businesses and goods are *not* fit for e-commerce? Why?
4. Find (on the Web) statistics about e-commerce. What conclusions can you derive from this data?
5. What added value do companies get from doing business over the Internet?
6. What properties of the Internet make it suitable for the e-commerce arena? Address properties such as interactivity, synchronism and asynchronism, and the Internet as a communication medium.

As mentioned previously, e-commerce encompasses many other topics: marketing, Web design, psychology, technology, cryptography, and more. E-commerce establishes a new economy with new rules. Here are two illustrative examples:

- In many e-commerce transactions, we process bytes instead of commodities.
- The value of online companies, which sometimes do not have any real estate, may be judged equal to the value of old and well-established firms.

Discussion

Andy Grove is the chairperson of Intel. One of his famous statements is "All companies must become Internet companies or die." (see *http://www.wired.com/news/business/0,1367,21849,00.html.*) Read this source (or other resources that refer to this vision). Suggest what reasons led Mr. Grove to make such a statement.

Almost any business can perform part of its activities in the online channel. Furthermore, there are services that emerged only as a result of the ability to offer online services and purchases. For example, several newspapers allow their readers to purchase selected articles and to pay only for these articles. This option is viable since the online channel enables this type of transaction and the online payment system can be adjusted for such situations by introducing, for example, concepts like micropayment.

Cognitive Analysis of E-Commerce

This section explores online purchasing by looking at two topics—metaphors and hypermedia—and investigating their influence on online buyers. The focus is on the end users.

TASK

In your opinion, what kind of customers does e-commerce especially fit? Address their characteristics, perceptions, and decision-making processes.

Metaphors in E-Commerce

In Chapter 2, "Software Engineering Methods," and in Chapter 11, "Abstraction and Other Heuristics of Software Development," we met the concept of metaphor. In all cases, metaphors are presented as a means of communication in software development. The same is true with respect to e-commerce. Metaphors are used in e-commerce environments to foster communication. The use of metaphors in ecommerce is almost transparent. Just think about the shopping cart that appears on many Web sites. Clearly, there are not physical shopping carts on e-commerce sites. The metaphor clarifies to shoppers how they should behave in these environments.

Let us put the discussion about metaphors in wider perspective by examining their contribution to learning processes. Constructivism is a cognitive framework that looks at the mental processes by which knowledge is constructed in learners' minds. Cognitive theories that look at mental processes from a constructivist perspective ask questions with respect to the mental *basis* on which knowledge is constructed. One way to provide learners with a basis for their mental constructions is by presenting metaphors that may help them relate new knowledge to mental structures that already exist in their minds. In other words, metaphors may provide a basis for the mental construction of new knowledge.

The same is true with respect to e-commerce. Since e-commerce is a new concept for all of us, the environments that support this commerce channel should attempt to associate the online shopping experience with customers' previous shopping experiences. More specifically, the role of metaphors in e-commerce environments is to help customers shop online and simultaneously associate their current purchasing actions with familiar ones.

TASK

Examine different e-commerce Web sites. List metaphors used to support one's orientation on these Web sites. In what way do these metaphors achieve their aim?

Hypertext and E-Commerce

Generally speaking, every piece of information can be presented in different ways. One might present a specific piece of information in a linear fashion while another person would use hypermedia. Cognitive theories tell us that our mind is a nonlinear "structure" [Minsky85]. More specifically, learners' (in our case, the customer's) knowledge is organized neither linearly nor in a flat structure. Thus, when we come to present information (to online customers), we should consider the fitness of hypermedia presentation styles, and *not* automatically adopt a linear fashion.

From a cognitive perspective, a hypermedia presentation of information supports the individuality of learners and the importance of letting them approach information in their own way. Jonssen [Jonssen86] suggests that because each individual's knowledge structure is unique, the sequence in which any piece of information is decoded by learners should be malleable, rather than set (p. 270). Thus, by allowing learners to choose the path in which they learn a certain topic (in our case, exploring an e-commerce environment), we give them first an opportunity to be guided by both their previous knowledge and their current needs and second, the choice of path to navigate, allocate information, read other customers' reviews, and perform other e-commerce activities.

The dominance of hypermedia in e-commerce Web sites can be observed easily by looking at the structure of a typical e-commerce Web site. Thus, for example, when one enters a company's Web site, in most cases, one may observe at a glance what the company offers. Then, the deeper the customer enters the Web site's hypermedia levels, the more details are presented about different products (their functionality, price, etc.). These details should lead customers in their purchasing decision-making processes.

TASK

Choose several e-commerce Web sites. Map their hypermedia structure.

Social Perspective of E-Commerce

It seems that one of the dominant social factors in e-commerce is the online communication among different customers and between the customers and the shops. For example, in the customer-customer relationship there are Web sites that enable customers to present their personal review about different products; the customer-shop communication channel enables e-commerce companies to announce new products and sales in a very simple way. Furthermore, e-communication enables customers to provide online shops with immediate feedback about their products and marketing strategy, information that may immediately influence the business directions that a shop takes.

Naturally, different e-communication tools serve this communication, such as e-mail, listservers, and discussion groups. To illustrate this aspect, Table 15.2 maps, in a way similar to Table 15.1, different communication tools according their synchronization with respect to time and place.

TASK

1. Suggest purposes for which each type of communication tool is especially suited.

TABLE 15.2 Communication Means: Time and Place Synchronism

	Time	
Place	**Synchronous**	**Asynchronous**
Synchronous	Face-to-face conversation	Physical bulletin board
Asynchronous	Telephone conversation	E-mail

2. Netiquette is the ethics of using the Web in general and e-communication in particular. Suggest at least three rules that address customers and the way businesses communicate electronically with their customers, which in your opinion should be included in Netiquette.
3. Customers' discussion groups may serve as a marketing research tool. For example, analysis of what is written in these discussion groups may tell us about customers' preferences, needs, problems, expectations, and so forth.
   a. Analyze one such discussion group. Based on this analysis, suggest practical implications.
   b. Compare the type of data obtained from the analysis of online discussion groups and data gathered by other data collection tools, such as questionnaires (see Chapter 4).
4. What factors might influence customers' decisions whether to purchase online. Specifically, what might prevent people from shopping online? How does e-communication deal with such feelings?

Discussion

Mobile commerce is the term applied to online financial transactions using a mobile device. This commerce fashion enhances, of course, the developments of mobile communication. (*http://www.mastercardintl.com/newtechnology/mcommerce/whatis/*). In what ways is mobile commerce connected to e-commerce?

THE TIMELESS NATURE OF THE INTERNET

Toward the end of the last chapter (Chapter 19, "Additional Information on Resources Used in the Book"), which expands our coverage of resources, appears a section on Web sites, something that would not have been present even a few years ago.

The Internet is an enormous mass of information. Some of it is obsolete, such as old conference Web sites. We cannot guarantee that you will go to a particular site and not get an error message error message of some type, because the owner of that site decided to restrict it to members. This is tantamount to an author going to each library that bought a copy of his book and collecting it. That probably just wouldn't happen. Web sites, however, can be withdrawn with just a few keystrokes. This section explores the timeless nature of the World Wide Web (the Internet) and how it affects the human aspects of software engineering.

In gambling rooms in the United States, there are no clocks so the gamblers will not know how long they have stayed or whether it is day or night. On Web sites, there seems to be no time as well. You do not have to wait for "hours of operation." Information is available 24 hours a day, seven days a week (24/7).

This 24/7 availability is good for students, who seem to be using references from the Web ever more frequently. Some book authors, with prior experience, are citing it now [Petroski03]. However, one has to be careful with these Internet citations, since the sites may be opinionated, limited to origination by a single person with a computer that can act as a server. In other words, a particular site may contain false information.

If you do a search for this year's conference on some topic, you may uncover previous years' sites as well, and perhaps some years may be missing. It all depends if the owner of the site disabled the links to it. This is another way the Internet is timeless. Old conference hotels are still being advertised years after the conference has come and gone.

TASK

Perform a search for your favorite conference Web site. Is the current version near the top of your findings? Why or why not?

The large number of devices attached to the Web supports this timelessness. A typical engineer seems permanently attached to a cellular phone, a personal digital assistant (PDA), and a laptop computer, at least. All these devices are likely to be wireless, so the Web is everywhere one can obtain a signal.

This timelessness is also supported by businesses like hotels, airports, and coffee shops that offer access to the Internet via wireless hot spots. It is also normal now for schools and information technology (IT) organizations to be similarly equipped. Therefore, accessibility is more of a rule than an exception. Common on our campuses are students checking their electronic mail (e-mail) with an open laptop computer wirelessly while walking between classes, and others who tune out lecturers by searching the Web or doing e-mail during class.

This tendency to tune out has even spread to other organizations: the first few persons entering a room for a long meeting in industry search for an electrical plug for their laptop computers. Speakers or those running a meeting do not even take note of the row of open laptops being used by people who do not even look at the presenter because they are using the Web as well as (maybe) listening.

The work/life balance exemplified by the eXtreme Programming (XP) practice of a 40-hour work week [Beck00] is destroyed by this easy accessibility. First, this practice does not mean "work only 40 hours a week," but work a reasonable amount each day and then rest. It certainly does not imply "work all the time," even though you can do that now.

Let us take the example of an executive going from New York to Chicago for an afternoon meeting. The executive can check e-mail at the home office, answer telephone messages in the cab to the airport with a cellular telephone, use a hot spot at the airport to do some more work in New York, work offline in the airplane, get some lunch at Chicago Airport while answering more mail wirelessly, answer the rest of the telephone messages in the taxi to the meeting, be on the remote office Internet during the meeting, and answer any further telephone messages in the taxi to the airport. Since the executive recharges the battery of the laptop during the meeting, she can work on the airplane and is then only a taxi ride away from the home wireless network. The entire day, the executive was connected invisibly to work. Similarly, as described in the first section of this chapter, people can be connected to shopping, movies, and other leisure pursuits. There seem to be two sides to anything on the Net.

TASK

At this time, using your and your organization's facilities, how long per day can you be connected to the Internet? How?

Communication

One payoff of the tech gadgets and the Internet is that communication is more frequent, if not improved, and some things that were once difficult are made easy. As an example, take this book. It is a collaborative effort between an author in Israel and one in the United States. Seemingly constant e-mail with attachments for review added to the mere couple of days of face-to-face communication. There also was almost daily comment by the editor. Certainly, regular mail could have been used, but the slowness of what is often termed "snail mail" was palpable. We will leave it to the reader to decide if this is a positive case study of communication!

Another positive use of communication: Jim Tomayko was sitting in on a new course. When he had an idea, or saw something on which he wanted to comment, he e-mailed it to the teacher on a wireless PDA. That way, feedback is immediate.

A way that feedback is slower, but still all electronic, are distance education courses. These, and an increasing number of courses on campus, have a Web site. A few semesters ago, one of the authors offered a course in a relatively standard room. So many students came that later ones arriving seemed not to touch the floor. There also was a shortage of books to accompany this shortage of space. Getting a different room solved the space problem; the book problem was resolved by putting a large number of copies on reserve. Both solutions were broadcast via the course Web site a few hours after the first meeting. In this way, the class did not waste time at the next meeting following the instructor to the new room, and many students read the assignment. In the days before Web sites, no-one would know which reserve shelf to use (there are several libraries on campus) until the second meeting of the class.

Another form of communication made possible by the Web was demonstrated to the authors during a program committee meeting for a conference. The chairperson had assigned three reviewers to each submission. Some had not completed their tasks, so the chairperson assigned reviews to be done at the Web site in real time during the conference call, by which call the committee made its decisions. The Chair could see integrated reviews appear on the Web site as they were completed.

A negative side of the modern Internet is the prevalence and growth of spam (junk e-mail) and viruses. Either can bring a company to its knees, the first through simple volume, and the latter in various ways. In an e-mail pile of 100 messages, 10 might be "real." Virus messages usually contain an attachment, which, when opened, does its dirty work. The growth of spam and viruses has changed the way we do Internet business. Viruses are quarantined or deleted by security software. Spam is now the subject of "spam blockers." This has affected distance education (DE) directly.

DE was invented to serve primarily those who were working but wanted to pursue a degree. Normally, they do so without giving up their job. Some e-mail or other communication is done during working hours, usually by the instructor, who figures general mail will be seen by the class later. For privacy, however, the commercially bought Web engine that makes sites for courses usually sends e-mail without naming those in the course. These broadcast-type messages are picked up and deleted by some spam blocker software, forcing the instructor to change communication methods.

Other modifications to on-campus courses include eliminating special tools and providing collaborative software for teamwork. It is essentially a poor idea to make a DE version of a course with the same characteristics of onsite courses, because the special circumstances of distant learners are ignored. Therefore, lectures are not just streamed or downloaded, but both. The objective is maximum learner flexibility.

TASK

List the communications media that you would use in a DE course.

SUMMARY QUESTIONS

1. Discuss dangers and pitfalls in e-commerce.
2. Suggest criteria for criticizing information presented on the Web on an e-commerce Web site.
3. This question continues the task presented in Chapter 4 in which you were asked to list requirements for an application that supports online surveys.
   a. Expand the requirements list you constructed for the online survey so that it also includes a means for e-commerce.
   b. In what way might the addition of these requirements influence the development process of the online survey tool?

FOR FURTHER REVIEW

1. Construct a model for an online company according to your choice. Present it to your colleagues for discussion and reflection.
2. The establishment of an online company requires one to consider different topics than those addressed to start a regular company. Suggest different topics that should be considered in such situations. Examine the differences between such situations and the more traditional ones. Address different topics such as employees and the supply system.
3. There are several payments means that were born together with e-commerce such as cyber-cash, digital cash, digital wallet, and micro-payments. Learn about these concepts (you can look at one of the online glossaries of Internet terms). Explain what problems customers may face when they use these payment means.
4. As is stated in the introduction to this book, this is not a human-computer interaction (HCI) book. However, HCI cannot be ignored when e-commerce is discussed. It is clear that the way in which an e-commerce environment is designed has a significant influence on its selling potential. This question presents several HCI-related considerations to be aware of when one is involved in the creation of an online company:
   a. What objects should be put on a Web site of an online company? What user's activity should each of these objects support?
   b. What relationships might exist between the design of a Web site and what the Web site sells?
   c. In what way is the design of an online company connected to the diversity of online shoppers?

5. One use of e-commerce is Web-Based Instruction (WBI). According to Khan [Khan97], WBI is a hypermedia-based instructional program that uses the attributes and resources of the Web to create a meaningful learning environment that fosters and supports learning. We add that it should also support teaching processes. WBI is adopted by different educational frameworks.
   a. Find on the Web examples of online programs. Analyze similarities as well as differences among these learning environments.
   b. Based on the analysis you conducted in (a), formulate different models of WBI.
   c. Analyze, from both the teacher's and the student's perspective, similarities as well as differences between WBI and on-campus learning.
   d. In what ways are WBI Web sites different from other e-commerce companies?
   e. In modern learning theories, terms like the following are emphasized: *interaction*, *active learning*, *knowledge versus information*, *learner freedom*, *huge source of information*, and *asynchronous in time and place*. How are these terms applied on the WBI Web sites you have reviewed?
   f. Discuss connections between WBI and copyrights. Suggest ways to overcome conflicts between these two concepts.
6. In the Study Questions at the beginning of this chapter, you were asked to consider connections between e-commerce and the Code of Ethics of Software Engineering. Rethink this relationship.

REFERENCES AND ADDITIONAL RESOURCES

[Beck00] Beck, Kent, *Extreme Programming Explained*, Addison-Wesley, 2000.

[Khan97] Khan, Badrul H., *Web-Based Instruction*, Educational Technology Publications (1997).

[Jonssen86] Jonssen, D., "Hypertext Principles for Text and Course Design," *Educational Psychologist* 21(4), (1986): pp. 269–292.

[Minsky85] Minsky, Marvin, *The Society of Mind,* Simon and Schuster, 1985.

[Petroski03] Petroski, Henry, *Small Things Considered*, Knopf, 2003.

The Alertbox: *http://www.useit.com.*
Center for Research in Electronic Commerce: *http://crec.bus.utexas.edu/.*
Shapiro, Carl, and Varian, Hal. R., *Information Rules*, Harvard Business School Press, 1998.

Lawrence, E.; Corbitt, B.; Tidwell, A.; Fisher, J.; and Lawrence, J. R., *Internet Commerce—Digital Modes for Business*, John Wiley & Sons, 1998.

Dodgson, M., "Organizational Learning: A Review of Some Literature," *Organization Studies* 14(3), (1993): pp. 375–394.

Drucker, Peter Ferdinand, *Managing for the Future*, Truman Talley Books, 1992.

Lakoff, George, and Johnson, Mark, *Metaphors We Live By*, University of Chicago Press, 1980.

Wallace, Patricia, *The Internet in the Workplace*, Cambridge, UK, 2004.

Part V
Software Engineering Education

Chapter 16, Case Studies in Software Engineering
Chapter 17, Students' Summary Projects and Presentations
Chapter 18, Remarks about Software Engineering Education
Chapter 19, Additional Information on Resources Used in This Book

16 Case Studies in Software Engineering

In This Chapter

- Introduction
- Objectives
- Study Questions
- Relevance for Software Engineering
- Software Management
- Software Development Paradigm
- General Principles

INTRODUCTION

This chapter contains 16 case studies from the software development field. The particular cases were chosen to illustrate some of the human aspects of software engineering problems. Each case study consists of a set of background questions designed to focus the reader's attention on the point(s) of the case study, narrative, and further questions to apply knowledge gained. The cases are of three types: the first group illustrates software management problems, the second group contains one case for each process of the paradigm of software engineering described in Chapter 8, "The History of Software Engineering," and the third group focuses on several problems involved with software design, architecture, and testing.

OBJECTIVES

- Readers will become familiar with information managers usually have, and its effect on the process of software development.
- Readers will observe the steps of the software development paradigm by analyzing case studies.
- Readers will gain tools to apply to software design, architecture, and testing problems.

STUDY QUESTIONS

Before reading the case studies, and based on what you have read so far in this book, what, in your opinion, would be the most common theme of the management narratives? The life cycle stories? The design cases?

RELEVANCE FOR SOFTWARE ENGINEERING

Since each aspect of the case studies pertains to some problem in software engineering and its solution, relevance is evident; we chose the cases based on utility. One of the most accelerated ways to illustrate such points is the use of case studies [Linn and Clancy92]. Table 16.1 presents the structure of the chapter.

TABLE 16.1 The Case Studies Presented in This Chapter

Software Management	Software Development Paradigm	General Principles
Case 1: Overtime	Case 5: Specifying	Case 9: The Recycling Principle
Case 2: Schedule	Case 6: Designing	Case 10: Multiple Representations
Case 3: Getting New Business	Case 7: Coding	Case 11: Alternative Tasks
Case 4: Discovering Information	Case 8: Testing	Case 12: Reflection
		Case 13: Fingerprints
		Case 14: Divide and Conquer
		Case 15: "Persecution Complex"
		Case 16: Literacy

SOFTWARE MANAGEMENT

Case 1: Overtime

Background Questions
1. What might cause any communication problem that this team has?
2. What was Don's team doing?
3. What did the team do when they got behind schedule?

Don Smith made his way past the first few dozen cubicles of the zone called Software Engineering according to the sign hanging from the ceiling. He finally reached the area where his small team of five was located. Actually, the team was larger than five, but some members were at another site on the West Coast. Sam, senior engineer on Don's team, was already on the network connection, busily coding on the big computer out there.

One thing the two sites had in common was that they were producers of development tools for the huge complement of coders in the company. The particular thing Don and his team worked on was a set of small procedures based on a Unix toolset in a book. Essentially, they were coding the book.

The project was in trouble: meeting with the Quality Assurance engineer who would be doing the testing revealed how far they were from turning the product over to him. It was Don's first project, and the amount of work done off site threw his estimates into disarray.

The question was: how to regain the schedule? All were already working as fast as they could. Was it possible to make an extra hour or so per day a requirement?

Don weighed the options in detail: six weeks of overtime versus extending the delivery date. He realized that schedule slips were likely to be blamed on him. He decided that overtime was better, as he had never heard of anyone being blamed for *that*.

After giving a rousing "we are the only thing acting as an obstacle to increased productivity" speech, the team worked 50 to 60 hours a week for the next six weeks. The software was never used. Don got another team.

TASKS
1. What was the "real" problem in this case? Could it have been avoided? How, or why not?
2. What was the impact of Don's decision on the other workers?

Case 2: Schedule

Background Questions

1. Who was Norma?
2. When one part of a plant goes on strike, what happens to the other parts?
3. Why did the project get off schedule?

Carol Pantene drove into the big parking lot next to the assembly building. She found a spot, and walked across the road to the modernistic building that housed her team. The first floor of the Tempest-compliant structure was given over to highly classified projects. She was on the second floor. There were many cubicles on that floor, some as a rest area for engineers on a classified project in the non-air-conditioned, brutally hot hangars. One cubicle was bigger than the others were. It was occupied by Norma, the head of the project.

Norma had a single vanity: she was really proud of her schedule. She ought to be: after nearly three years, they were less than a week off. Carol had never been on such an on-time project.

The next morning, as Carol drove over a rise just south of the plant, she saw cars stretching along toward the entrances to all the parking lots. When she finally (after four hours) got to the head of the line, she saw people wearing signs. She could only make out "Machinists on Strike." That was enough. She figured out that the people blocking the way were some of thousands of striking machinists whose contract had run out. During the last machinists' strike, Carol and the other engineers were unaffected, as their biggest competitor's machinists went on strike. Since her company and its competitor were the last corporations left in the country that produced the artifacts in question, machinists first struck one, then the other, trading off on each contract, so as to not bring the industry to a complete halt. The affected company, left without any income, eventually settled the strike, and the working machinists at the other company quickly voted to accept the same terms.

Carol was nearly five hours late for work. She arrived just in time for upper management to send her and the other engineers home. Norma had lost a day on her schedule.

The next morning, the line was a little shorter. When she got to the head of it, only two strikers were at the entrance to the car park. Management had obtained a restraining order the night before, limiting the strikers to two persons at each entrance. The strikers were walking across the road like guards, so, for only a few seconds on each crossing they were far enough apart to enable an engineer's car to pass without running them down. The flow inbound was jerky but steady. She was only three hours late this time. Others were there ahead of her, but some came in during the next hour. Between late arrivals, and general disruption, Norma lost another day.

The next morning, the line was almost as long as the first day, but it did not seem to be moving. As she neared the level approach to the plant, some people waved their arms for her to stop her car. Apparently, some enterprising young machinist had filled his pickup truck's bed with roofing nails. These had been spread on the approach road during the night. Cars further down the road had four flat tires. Obviously, no one got in that morning. Norma's team lost another day.

Over the three weeks of the strike, about a week was wasted, doubling Norma's scheduling error. She was upset.

TASKS

1. What was the nature of the strike risk (where did it come from)?
2. What could Norma have done to mitigate the strike risk?
3. How did the strike affect the project work?

Case 3: Getting New Business

Background Questions

1. How did the team devise the original estimate do its job?
2. Why did the senior managers change the schedule?

Mike Culpepper sat hunched down, trying to be "invisible" near the back of the auditorium. At the front were managers well above his pay grade. It was obvious that his company had to diversify. For 40 years they built big airplanes, but now no or few big airplanes would be bought. They had decided to move into the tactical airplane business, something that they had not built for nearly 50 years. This would be the first contract on which they would bid in the new arena.

Mike and his coworkers had concentrated on a Request for Proposal (RFP) to upgrade existing but no longer manufactured tactical airplanes for a state national guard. Knowing how important this was for the company, his team spent much care with the estimates and especially the bottom line. This day was a review of their plans before shipping the proposal to the customer.

The Grand Pooh-Bahs sat near the front, took everything in, and then, after Mike's team was done, announced, "We can't win at this price," and detailed a new figure. They won the contract as low bidders.

The team estimated the number of workers needed for each phase of the development cycle. Since they were using something that looked like the waterfall model, they expected to start small, ramp up, and go back to being small as the deadline approached. The new price meant that the height of "ramping up" was now less than before. This meant fewer workers at key times. All was all right in the first phases. When the team entered code and test, being shorthanded started to hurt. As they went along, the number of engineers assigned to them got even smaller. The managers claimed that the extra help had worked too long on the project, and was no longer part of the budget.

Mike's team reacted by beginning to slip milestones. One by one, people were being transferred according to the original schedule. Slowly the team's cubicles emptied out, until only Mike remained for the last touches on the software. He delivered it two years late.

TASKS

1. Whose fault is it that the project was late?
2. How could lateness been avoided?
3. Should Mike's team have taken the contract?

Case 4: Discovering Information

Background Questions

1. What was the army of accountants trying to calculate?
2. Is this information useful to software development teams? Why, or why not?

John Nunez was taking a shortcut using the administrative building added to the government-built plant. He was going back to his cubicle office after checking on some assemblies. It was toward the end of the day.

John crossed the open-space work area (all desks, without walls). He walked his normal "engineer thinking" walk: hands clasped behind his back, head down. As he passed one desk, the words "lines of code" leapt at him. He scrambled back to the desk, startling its owner. What John found amazed him.

Each engineer filled out a card every day. This was done so the company knew how much to charge the government. They entered lines of code produced and time. Thus, they had an idea of personal productivity.

Here, the accounting clerk separated and summed the data by contract number, thus deriving the corporate productivity. "How long have you been doing this?" John asked. "For years," the accountant replied.

John remembered his boss moving from her old cubicle to a new one, carefully cradling copies of all her written estimates on every previous project. He thought about how much real data by project would have helped her. He also thought about how much real data for all the projects would have helped the company.

TASKS

1. What risk does a lack of real data cause?
2. Is there a way to address this with the accounting data?
3. What do you think of a big company acting this way?

SOFTWARE DEVELOPMENT PARADIGM

Case 5: Specifying

Background Questions

1. What is the functional specification used for?
2. Why were there errors in it?

Jane Doremus found about 100 pages of printed sentences with lots of white space on the center of her desk. It was the "Functional Specification" of the software product for which she had to do the user manual. This was to be the source of information that Jane was to use for composing the document.

Unfortunately, the specification constantly changed. The usual method was to write the manual using the spec as the best source, and then the manual was given to the technical lead of the project for review. She marked it up, based on her understanding of the requirements to date.

Jane was writing the manual. While she did so, she realized that only rarely did her work match the software when it was done. The result was a flurry of bug reports.

TASKS

1. Who really found discrepancies between the user manual and the software?
2. What do you think about this way of writing user manual?

Case 6: Designing

Background Questions

1. What is the difference between "design" and "architecture?"
2. Why did Jill let the design document fall out of currency?

Jill Smith was known as the System Architect. She looked carefully at the detailed design to date. Once past this point, the software would be outsourced to the company's partner in China.

Usually, the partner would get a current description of any interfaces for software they would add. Two things convinced Jill to skip the preparation of this document. One was that the project was late. The second was that she saw a Chinese man while visiting the company who looked to be from the mainland of China, where the partner is located.

TASKS

1. Was the system architecture ready to be outsourced?
2. What were Jill's problems? How could she have avoided them?

Case 7: Coding

Background Questions

1. How do the architects handle detailed design?
2. How did Jim violate the design?

The detailed design filtered down from the team of architects above. Jim Royce got his copy of his component early Tuesday morning. Quickly, he coded the series of equations and the interfaces to his module.

Friday at noon he was cursing the stupidity of the architects. He could clearly see how the exchange of data could be optimized. They had missed it.

He could file a change request, but everything else was going well, it was a small change, and only one other engineer was affected. Jim made the change himself. Later, he sought out the affected programmer and told her that there was a change.

TASKS

1. What is the result of Jim's lying to the other engineer?
2. Does optimization matter here?

Case 8: Testing

Background Questions

1. What was Nian trying to do?
2. How did she do it?

Qu Nian entered the collection of modules into her automated tester. She could almost hear it chugging along, examining every module. It tested sending messages and accurately receiving them.

Even with the speed of automated testing, Nian was certain that she could not be done before the end of the day. She decided instead to make a more limited test suite. This one exercised only a portion of the software, saving considerable time. She figured that if she used Markov modeling, only a portion of the test suite would be run, anyway.

TASKS

1. Why was Nian's reasoning false?
2. What else could she have done?

GENERAL PRINCIPLES

Case 9: The Recycling Principle

Background Questions

1. How could Joe reduce his work?
2. What still needs to be done?

Joe Johnson was convinced that he had seen this design before. He knew that he was assigned to the project team as the domain expert and, incidentally, architect. He certainly never expected to become a software architect.

He had used the architecture on the white board before. In fact, it was right out of [Garlan and Shaw96], which he kept on his shelves. He looked at it now, seeking confirmation. Sure enough, there it was. Joe realized that if the domain was the same, he did not have to take a totally fresh look at the problem to design a solution. Rather than reinventing the wheel, he could use an existing design that had been available all the time.

TASKS

1. How is the Recycling Principle evident here?
2. Why and how does a human aspect of software engineering fit into this case?

Case 10: Multiple Representations

Background Questions

1. What was Jason charged to do?
2. What did he have available to him to do it?

Jason Smith examined the five "views" of the software product. He intended to pull architecture from these UML constructs. In fact, every representation of software that he ever worked with had multiple insights useful in devising architecture. It all started when he read Kruchten's "4 + 1" article [Kruchten95]. The uses of multiple ways of looking at things are a staple of UML, and he was trying to bring the idea to his group.

TASKS

1. What practices go into building architecture?
2. Why did Jason center on UML?

Case 11: Alternative Tasks

Background Questions

1. What did Barbara consider when she was trying to find the best design?
2. Why did she have these alternatives?

Barbara Count knew that there were more ways to skin a cat. She could see several ways to solve the software problem. Rather than centering on just one, she worked out several designs based on her models. Then she evaluated the designs.

Weighing quality attributes against functional attributes, she looked for the best compromise. Herb Simon once called this "satisficing" [Simon69]. Finally, one design emerged from her pile.

TASKS

1. Why did Barbara have to look at multiple designs?
2. Why is this extra step possibly advantageous?

Case 12: Reflection

Background Questions

1. What did the engineer examine?
2. How did she do it?

Yael Levi balked at doing tasks she felt were unnecessary. She had vague feeling that some things required by the government's software development process were an advanced form of busywork. Some of her coworkers joked about the 18-wheeler truck that would be needed to hold all the documentation.

Yael looked at each document and reflected on how her project used it. Some were unused, some were used multiple times. She was able to go to her team leader with the results of her reflection, a list of the unused documents, thus giving her team ammunition for waivers from the customer. Now she began to look for wasted effort in the code.

TASKS

1. How might Yael's reflection help her team leader?
2. What does "reflection" mean here?

Case 13: Fingerprints

Background Questions

1. How did Sam find the defects in his code?
2. Why was the code difficult to understand?

Sam Jones was debugging a chunk of code produced by the Unified Process. He was at first bewildered. The code seemed too intractable.

Then he listed things that looked like bugs he had fixed previously. Before running any more tests, he made up a suite based on his memories of recent bug finds. Running this test suite, looking for bugs with similar "fingerprints," he was able to gain a lot of familiarity with the defects in the code. The software became understandable, and Sam felt better.

TASKS

1. What are "fingerprints?"
2. How do such symptoms help to find bugs?

Case 14: Divide and Conquer

Background Questions

1. Why did the large group meet?
2. How were the components limited?

All the software engineers gathered in one big auditorium. Upper management introduced the new product. At first, it seemed like "code the world." There appeared to be no front or back to the software, and it was complex.

Dan's team gathered the next morning with a single component to code. After they finished that component, there was another. Every component was smaller than 150,000 lines of code. Most were less than 100,000. He could see how the components fit together by thinking back to the first meeting. This kept his thinking more in order.

TASKS

1. Why were the components kept relatively small?
2. Why did Dan feel better about the project?

Case 15: Finding Hidden Bugs

Background Questions

1. What is "white box" testing?
2. How does it miss defects?

Sarah Rehme looked at the pile of printouts for the code she was testing. Somebody else had already done white-box style path testing. She had nearly "clean" code.

So, the question was, how do you find the remaining bugs? She made up a test suite of possible weaknesses and extreme cases. These she ran, looking for the hidden defects by identifying bugs that were found by these constructs before.

TASKS

1. How does Sarah's test suite represent a persecution complex?
2. Can you think of another test suite theme?

Case 16: Literacy

Background Questions

1. What do you think made the code easy to read?
2. Why was finding a place to put the new code easier?

Jo was surprised at how easy the code was to read. It was sparsely commented, but, unlike other code she had tried to read that had few comments, it flowed like a romance novel. She read it with pleasure.

A maintenance job came down the pipeline. Jo was able to find a logical insertion point very quickly. She added the code and passed it to Sarah for regression tests.

TASKS

1. What are regression tests?
2. Why are human beings happy to see such code?

FOR FURTHER REVIEW

1. Make a chart, by topic, listing ideas gleaned from the case studies. To what topics discussed in the book are these topics connected?
2. In the next chapter, Chapter 17, which presents summary projects and presentations, you are invited to develop and present case studies. As a preparation, list features that in your opinion a case study should have in order to achieve its targets.

REFERENCES

[Garlan and Shaw96] Garlan, David, and Shaw, Mary, *Software Architecture: Perspectives on an Emerging Discipline,* Upper Saddle River, NJ, 1996.

[Kruchten95] Kruchten, Phillippe, "Architectural Blueprints—The 4 + 1 View Model of Software Architecture," *IEEE Software,* (November, 1995): pp. 42–50.

[Linn and Clancy92] Linn, Marcia C., and Clancy, Michael J., "The Case for Case Studies of Programming Problems," *Communications of the ACM* (March 1992): pp. 121–132.

[Simon69] Simon, Herbert, *Sciences of the Artificial,* Cambridge, MA, 1969.

17 Students' Summary Projects and Presentations

In This Chapter

- Introduction
- Objectives
- Study Questions
- Relevance for Software Engineering
- Case Studies
- Construction of Case Studies
- Presentation of Case Studies

INTRODUCTION

This chapter presents several ideas to be used by "Human Aspects of Software Engineering" course instructors as the end-of-semester projects. Specifically, we suggest three directions: a fictional case study construction and analysis, a report about a field study conducted by the students, and the description of an event that happened in a software production house in the past. In the first case, learners are asked to develop a case study and to analyze it from the different perspectives discussed in the book. In the field studies, learners are asked to visit a software development house, to observe the work of software teams, to focus on one interesting phenomenon, and to analyze it. In the description of a historical event that has happened in a software house, students are asked to identify such an event and to analyze it in a manner similar to the other two tasks. In all these cases, students' analysis should be multifaceted and address different topics presented in the book. The main difference between the tasks is shown is their creation process. While the first is a more theoretical process, the second and the third are based on the observation of software teamwork and the analysis of other artifacts.

OBJECTIVES

- Readers will become familiar with characteristics of case studies.
- Readers will become familiar with processes of case study development.
- Readers will identify key issues to be addressed in case studies.

STUDY QUESTIONS

1. What is a case study?
2. In what fields are case studies used? What purposes do case studies serve in each field?
3. Assume you are a team leader and a conflict is identified between two members of your team. You do not want to take a side, but you do want to discuss this issue with your team. Select a topic that may cause a conflict in software teamwork, and describe a situation in which such a conflict may be raised. How can a case study be useful in such situations?
4. Suggest additional situations in which a team leader might choose a case study as a management tool.
5. In Chapter 5, "The Code of Ethics of Software Engineering," we used many case studies to illustrate the essence of the code and its use. In your opinion, why is the topic of ethics appropriate to be discussed using case studies? Suggest additional topics that in your opinion are appropriate to be discussed through case studies.
6. Case studies can be viewed as storytelling. Indeed, storytelling is a management tool. There are those who argue that storytelling will become a key ingredient in management, education, training, and innovation in the twenty-first century (*http://www.creatingthe21stcentury.org/*). Explore the Web and find information about the use of storytelling as a management tool. Sum up your conclusions.

RELEVANCE FOR SOFTWARE ENGINEERING

Case studies are a useful way to communicate ideas, especially in cases in which one introduces abstract ideas. Case studies enable one to present the essence of the idea, yet not to present it too abstractly. In this chapter, you will be asked to construct, relate, and analyze case studies. The relevance to software engineering is clear. First, in the process of building the case studies you will address software engineering issues. Second, when you analyze your case study, you will deepen your understanding of the relative importance of the different topics mentioned in the case study. Such understanding may improve your understanding of software development processes in general.

CASE STUDIES

Case studies focus on situations related to a specific profession. In most cases, they are presented as stories, with actors and protagonists. In this way, case studies enable us to examine concrete details prior to or after the examination of an abstract theory. Each case study should have point(s) it aims to highlight and increase readers' attention to. Generally speaking, if a case study brings to the surface the topics it aims to highlight and leads its readers to think on and analyze these topics, we may say that the case study achieves its goals.

At the end of Chapter 16, "Case Studies of Software Engineering," you were asked to identify features of case studies. You are invited to add your characteristics to our suggested list of case study traits:

- Focused
- Vivid
- Relevant for the audience
- Not trivial
- Highlights known topics from new perspectives

Case studies can be used in different ways. For example, a lecturer can ask a class of students to discuss a case study in groups and then present their conclusions in front of the class. After the different groups present their analysis and opinions, these approaches can be discussed, clarified, and elaborated. Another way in which case studies can be used is in job interviews to learn about candidates' attitudes toward specific situations that might be raised in work. In such interviews, the interviewees can be asked to analyze case studies presented to them. The attitude reflected in their analysis may be one of the hiring factors. Chapter 16 exhibits another way by which case studies can be used. It focuses on specific events in software development and highlights them from the different perspectives discussed throughout this book. The case studies presented in Chapter 16 are short in order to concentrate readers' attention on specific topics.

Case studies are used in many professions, such as business administration, medicine, and law. In all these fields, the purpose of using case studies is similar.

The differences are expressed, of course, in the contents of the case studies and the terminology they use. Each community adopts a slightly different writing style in presenting its case studies.

CONSTRUCTION OF CASE STUDIES

Chapter 16 presents 16 case studies, some quite short and limited. In this chapter, readers are asked to construct case studies. The process of constructing case studies can be carried out in different ways. One way is to visit a software house for a period of time, identify an interesting event, and report about it in the case study. Another approach is theoretical: the case study is constructed based on a fictional, theoretical analysis of the topic one wants to explore. Another option is to report on an event that happened in the past and tell it as a story. All these options are described in what follows. The described processes are our suggested procedures, and you may alter them if you find it appropriate. As long as one produces a reliable case study that raises stimulating questions for discussion from which the readers can learn, the process of case study construction is of less importance.

Option 1: Construction of a Theoretical Case Study

This option lays out a process for the development of a case study based on the analysis of topics introduced so far in this book. The aim of this process is to highlight the immediate application of the theoretical knowledge described in the book to the real world of software development. Personal experience has an important role in this process as well and should not be ignored. This experience is reflected in the assessment of what topics are relevant, in decisions about what topics should be left in the case study, and in other similar situations.

Our recommended process for the case study construction is composed of six steps. After the process is outlined, it is demonstrated by the construction of a specific case study.

Step 1. Select a topic: Think about a topic you find interesting and relevant for you to discuss.

Step 2. Analyze the nature of the topic: In this stage you are asked to check whether the topic you want to focus on has enough heft to be at the center of a case study. Ask yourself questions such as:
- To what software development activities is the selected topic connected?
- To what players in software development environments is the topic connected?

- What human aspects of software engineering does the topic address?
- Is the topic connected to the individuals of the team or is it connected to the team?

If your answers to the preceding questions indicate that the topic is indeed rich enough and can be connected to different issues in software development environments, it might fit to be a central topic of a case study.

Step 3. Imagine possible situations: Imagine at least two situations in software engineering in which the topic may be relevant. The idea is to check whether there are specific situations in software engineering in which the topic you want to pursue has a significant expression.

Step 4. Write the case study: Start writing down the selected case study. Try to make it as vivid as possible without forgetting to include in it the main issues you wanted to address.

Step 5. Check the scope of the case study: After you complete the first writing (and editing) of the case study, check whether other related topics can be added to the case. Make sure not to alter the focus of the case study. Then, check issues such as: Is your main message conveyed properly? Are the connections between the different topics addressed in the case study clear?

Step 6. Develop questions about the case study: Develop stimulating questions that are appropriate to be explored with respect to the case study you just developed.

Illustration of the Process: Developing a Case Study

Step 1. Select a topic: In this first stage, you are asked to recognize an interesting topic for you to discuss. In our case, an interesting topic is a topic related to software engineering whose human aspects may raise interesting questions, dilemmas, and solutions. In fact, if you just think openly enough you will find that each topic related to software development may have interesting human aspects. For example, think about IDEs (Integrated Development Environments). While the actual installation process of an IDE may be conceived of as a technical issue, if we expand our view we find that IDEs may raise very interesting human-related topics. Here are several human-related questions about IDEs: How should a team select the IDE with which to work? What features of an IDE are more important for a specific development process? A specific project? A specific team?

Following this short analysis we can get the feeling that the concept of an IDE may have the potential to be an appropriate topic to construct a case study around. Consequently, it will be the focus of the case study that we will develop in the next stages.

Step 2. Analyze the nature of the topic: Here are partial answers to the questions we suggest to address to check the fitness of the topic of IDE to be at the center of a case study.

- IDEs are connected to many of the central activities of software development. While good IDEs support the development process, bad IDEs may interfere with development processes. Accordingly, a selection of one IDE over another one may have a significant influence on a software projects' success.
- IDEs are connected to the main players involved in software development processes. Because the IDE is the environment in which the code is developed, coordinated, and integrated, it has a significant influence on teamwork, customer's acceptance tests, and more.
- Since IDEs are connected to many of the development process activities, they are naturally related to different human aspects of software engineering. Just think about the potential influence of an IDE that supports refactoring on development processes.
- Features of an IDE have a direct influence on the level of teamwork it supports. Thus, the topic of IDEs is connected to software engineering processes, both on the team level and on the individual level.

Step 3. Imagine possible situations: Here are two software engineering situations in which IDEs may have significant expression.

- A decision process about what IDE to adopt for a specific project. A case study that addresses questions like the following can be built:
  - **Evaluation of IDEs:** What considerations should be taken into account in the process of selecting an IDE?
  - **The inclusion of specific menus, such as testing and refactoring menus:** Developers may feel that if a specific IDE, which contains these features, is selected for their project, they will have to apply these activities and may exhibit resistance.
  - **Cost versus quality in the assessment of an IDE:** If a choice has to be made between an expensive IDE of high quality and a freeware IDE of a lower quality, what considerations should be counted?
- **The development process supported by an IDE:** Such a case study may address topics such as what software development methods the

IDE should support and what main features it should include. With respect to each topic, the case study can present different opinions, suggested by different people in the organization, that stem from different objectives and motives.

Step 4. Write the case study: Start writing down the case study. Try to make it as vivid as much as possible but make sure it is realistic. In what follows, we present one possible case study that focuses on IDEs.

A Case Study about IDEs

IDEIDE is a large software development house, with 400 software engineers and about 100 marketing and administrative employees. The product that gives IDEIDE its prestige and its competitive advantage is a software tool that guides software developers in testing processes.

It was a spring day when one of the senior team leaders, Gil, heard about eXtreme Programming (XP) and wanted to check its fitness for her team. Looking around for the best consultant XP company, she found XPTruth to be the perfect consulting company to introduce XP into IDEIDE. It was clear that the process of introducing XP into IDEIDE would not be smooth since the process by which software was developed in IDEIDE at that time was significantly different from the one that XP advocates. For example, IDEIDE developers used to add and develop features that they had not been asked to develop, developers did not perform unit tests (the QA unit performed the testing), and integration was conducted only after the development processes of several parts of the developed software had been completed.

XPTruth is a two-person consulting company that introduces XP to companies by presenting and illustrating the XP values and practices. In each presentation that the XPTruth team performs, it lets the audience experience the essence of XP as much as possible. Each introduction of XP is summed up by discussing with the audience the fitness of XP to their particular company.

The XPTruth team accepted Gil's invitation to introduce XP to a group of team leaders at IDEIDE. As in other cases, at the end of the presentation, a discussion about the fitness of XP to IDEIDE took place. This discussion raised many interesting and unexpected issues.

Unexpectedly, the team leaders did not talk about the fitness of XP for their company, but rather, they started discussing the testing tool that IDEIDE develops. It seemed like they totally forgot about the attendance of the XPTruth consulting team. The discussion emerged as a response to a short illustration of JUnit, which is a testing framework written by Erich Gamma and Kent Beck, used by developers who implement unit tests in Java (*www.junit.org/index.htm*).

The brief JUnit demo raised questions related to the testing software developed by IDEIDE. Among the main questions, the following were raised: In what sense is the JUnit framework different from the testing framework that IDEIDE sells? Should IDEIDE adopt the test-driven development (TDD) approach? At a specific stage, Dan, one of the respected team leaders, raised the following question, which left everyone surprised: Are IDEIDE products tested properly? Don't they contain bugs that could have been avoided if IDEIDE had worked according to the XP practice of TDD? The last question is critical, as IDEIDE is a software house that aims at producing a product that supports software-testing processes.

After about an hour of discussion, during which the XPTruth consultants just listened to the give and take, they were invited to address this topic. To pursue their message, the consultants illustrated an IDE that uses JUnit as its testing framework. The message was clear.

After the lunch break, during which the team leaders had a chance to digest the facts that had just surfaced, the discussion continued. The main topics that were addressed were: Should IDEIDE products include JUnit? Should IDEIDE develop a similar testing framework? Should IDEIDE develop its own products by using JUnit? If defects in IDEIDE's delivered products are found by inspection, how will it inform its customers about them?

Each of these questions led to the expression of many emotions. Some of the team leaders began to be frightened about losing their jobs. Others not only were shamed by the way they developed software, but also started feeling badly that they would have to tell their customers about the defective product that they delivered for so long. IDEIDE's CEO, who attended the meeting, started thinking about the financial consequences of each of the possible answers to the previous questions.

Step 5. Check the scope of the case study: At this stage, after we have completed the composition of the case study, we should check whether other related topics can be added to the case and whether the main message we wanted to deliver by this case study is reflected properly.

Our rereading of the case study reveals that IDEs play a central role in it. IDEs are addressed through software developers' attitudes with respect to the adoption of IDEs and the features that IDEs should offer. Here are two additional topics we could have addressed at this stage:

- **Refactoring menu:** Some IDEs contain a refactoring menu (see Chapter 9, "Program Comprehension, Code Inspections, and Refactoring"). Such a tool enables the performance of many refactoring actions more efficiently and more conceptually. More specifically, the refactoring menu enables one to think about the design aspects of each refactoring move instead of being swamped with the details and the technical aspects of the specific refactoring action. After rereading the case study, we decided not to add the topic of refactoring to avoid excess complexity. What we decided to do is to address refactoring in the questions presented to the readership of the next case study.

- **Budget:** Budget considerations are a central topic in the decisions that the team leaders of IDEIDE will have to make. If you look at the last sentence of the case study, you will see that it addresses this topic. Indeed, this sentence was added only as a result of this step in the process of the case study construction.

Step 6. Develop questions about the case study: At this stage you should develop stimulating questions to be explored with respect to the case study you just developed. Our analysis of the IDEIDE case study revealed that it addresses the following list of topics, with respect to which we can present questions: software development methods, software project management, testing, ethics, decision making, and software teamwork. Among many possible questions that we could present, we chose the following:

1) Imagine that you attended the meeting described in the IDEIDE case study. In your opinion, what specific characteristics of the development process that is currently used in IDEIDE led to the emergence of each of the questions mentioned in the case study?
2) Assume that IDEIDE decided to initiate a process in which it will start checking its products thoroughly. That checking revealed that IDEIDE's products for guiding testing processes contain bugs! In your opinion, what approach should IDEIDE take toward telling its customers about this fact? What does the Code of Ethics of Software Engineering say about such cases?
3) What information do you need to determine whether IDEIDE products are failures or successes as software products? How will you collect this information?
4) Refactoring is an integral part of some of the newer IDEs. What considerations, in your opinion, led the developers of these IDEs to add a refactoring menu to the developed IDE?

So far, we illustrated in detail the process of constructing a case study based on our theoretical analysis of what topics are appropriate to be introduced into the case study, the translation of these topics into human behavior, and their description as a story. In what follows, we present two additional ways by which one can write a case study. These processes start when one visits a company in order to find an interesting story to construct a case study based on it. When composing a case study based on a real event that happens in a software house, you can concentrate either on an event that happened in the past or on a case that happened during your visit to the company. In what follows, we briefly explain the main stages of constructing each type of case study.

Option 2: Construction of a Case Study Based on a Field Study

When constructing a case study based on a field study during a visit to a software house, you should feel like an anthropologist who enters a new society and looks for interesting behaviors to report about. In such cases, one cannot decide ahead of time what the topic of the case study will be; rather, one must arrive with an open mind and start looking for interesting topics that may have the potential to become the heart of a case study.

For describing a case study based on what happens during one's visit to the development site, one should be very sensitive to what goes on. The best way to get such information is to conduct observations (see Chapter 4, "Software as a Product").

Before starting any such observation, you should get the permission of the relevant authorities to conduct these observations. In addition, you should inform the people you are going to observe that they are going to be observed. It is very important to explain to them why you will observe them and what specifically you are going to observe (concentrate only on software development-related issues!). In addition, promise these people that the observations will be used only for the description of the case study, and that their identification will be kept anonymous.

When you start the observations, try to be passive and unseen as much as possible. Such behavior will increase people's trust in your presence. Do not get involved in their daily work. Just keep writing and documenting your observations. When you notice something interesting going on, try to locate all the events that seem to be related to it. By "all events" we mean all the discussions, remarks, changes in plans, and so forth that seem relevant to your case study. Close attention to details will enable you to report about the event more vividly.

After the completion of the observations, when you feel that you have enough data, start describing the case study. When it is relevant, also describe the development environment, tell about the setting of the development site, and add details such as the number of people involved, their professional background, and any additional details that will enable the readers to construct in their mind a picture of the case study.

After you complete writing the case study, ask the people who participated in the case study to read it and to comment. Their input can be very useful for the case study's improvement. For example, they may explain their own as well as the others' behavior. When they react, do not criticize or argue with them. Recall that they are sharing their interpretation. You should listen, try to understand their perspective, and check whether you can use their input for the improvement of your case study.

After you incorporate the participants' observations into the case study, read it with fresh eyes and develop relevant questions.

Option 3: Construction of a Case Study Based on an Event in the Past

In the case of describing an event that happened in the past in the software house you are visiting, there are two options. In the first option, you may know about an event that happened in the past. In the second, you must identify an event. In the second case, you can start with interviewing people in the organization to find out whether there is an event that seems to be central in the collective memory of the organization (see Chapter 10, "Learning Processes in Software Engineering"). After such an event is recognized, one can continue as described here with respect to the case of a known event.

To describe a case study that happens in a specific company in the past, the best way is to base the case study on the information obtained from different research tools, such as interviews, questionnaires, and related documents. For additional information about these research tools, see Chapter 4, "Software as a Product."

Here we briefly elaborate on document analysis, a topic that we did not address in detail in Chapter 4. Documents are written artifacts, like minutes of meetings, discussion groups, procedures used in the company, and any other material that people write during their work. The analysis of documents can be done on different levels. First, one can refer to the chronological details of the described event. Second, one can interpret the hinted tone of what is written, trying to speculate on the motives that led someone to write or say what is written. These interpretations should be validated in later stages with the person who wrote the document and with the quoted people. Third, one can try to connect these interpretations to the atmosphere and culture that characterize the described development site. These speculations should also be validated with the relevant people.

As previously mentioned, after you complete the writing of the case study, it is important to give the involved people time to read it and comment. In this case, such feedback has at least two purposes. First, you will be able to validate your interpretation of the gathered data. Second, these people may shed additional light on what has happened. When they read the case study, they may recall additional relevant details.

After the completion of the validation stage and the addition of the relevant details, you can move to the development of questions.

PRESENTATION OF CASE STUDIES

After you finish writing the case study and the questions, the time arrives to present your case study in class. Here is a recommended format. It is illustrated using the IDEIDE case study:

- Start by presenting your classmates with a general question related to the case study. For example, "Does anyone work with an IDE? Can you tell us your impressions? Benefits of IDEs? Pitfalls?" Dedicate to this stage only a few minutes. The idea is to get a brief impression about what your classmates know about your topic.
- Tell the case study. If possible, connect the case study to what your classmates have said before.
- Present your classmates the questions you developed about the case study. When they discuss these questions, try to talk as little as possible and let them talk as much as possible. Invite them to respond to each other's opinions.
- Summarize on the board the main points raised by your teammates.
- Ask your classmates to propose additional questions about the case study.
- Discuss your classmates' questions as well.
- Summarize the entire discussion with a few sentences that capture your perspective of the case study, taking into account the audience's expressed opinions.

Discussion

After all the case studies are presented in class, it is worth discussing with students both the process of constructing the case studies and the lessons learned from the class presentations and the discussions that followed them. Each instructor can find his own relevant structure for such a summary lesson.

ADDITIONAL RESOURCES

Storytelling *http://www.creatingthe21stcentury.org/*
Engineering Case Studies *http://www.civeng.carleton.ca/ECL/*

18 Remarks about Software Engineering Education

In This Chapter

- Introduction
- Objectives
- Study Questions
- Relevance for Software Engineering
- The History of Software Engineering Education
- The Education of Software Engineers Today
- Teaching Human Aspects of Software Engineering

INTRODUCTION

In the past couple of years, terms such as *chasm* and *crossroads* have been heard in discussions about computer science and software engineering in general and their education in particular. Here are several examples: Glass's paper "Revisiting the Industry/Academe Communication Chasm" [Glass97]; El-Kadi's paper "Stop That Divorce!" [El-Kadi99], which urges us to stop the divorce between computing and software engineering since there is too much at stake (p. 28); and Gudivada's paper [Gudivada03] "The Computing Profession at the Crossroads," which calls industry and academia to work together to chart the best course for the profession's future (pp. 90–92).

In parallel (and perhaps as a result of these voices), educators suggest exploring new directions for the education of future software engineers. Among others, Denning in his "Educating a New Engineer" article [Denning92] suggests that if a curriculum wants to prepare students for a changing world, it must incorporate new elements, such as effective interaction with others and a greater sensitivity to the historical and cultural spaces in which we live and work. [Yeh02] says that in addition to the traditional core study of computer science and engineering education, we must also integrate into the software engineering curriculum the basic knowledge of business performance and measurements, fundamental skills of communication, and human problem solving in team environments.

Each of the aforementioned articles emphasizes a different gap or elements to be added to the software-engineering curriculum. Some refer to the lack of human skills in the training of new software engineers and others suggest strengthening the mutual responsibility of industry and academia when a new curriculum for software engineering is constructed. One common idea stands behind all these calls for action, however: all suggest broadening the education of software engineers beyond the scientific theoretical courses. This book about human aspects of software engineering is written in that spirit. We hope that it will be useful in computer science and software engineering programs.

OBJECTIVES

- Readers will conceive of the importance of the education of software engineers and its direct influence on the way software engineers perceive their profession.
- Readers will be able to identify the strengths and weaknesses of their professional education.
- Readers will be able to identify what they should learn to improve their professional skills.
- Readers will form a conception of the role and importance of a course that deals with human aspects of software engineering in software engineering programs.

STUDY QUESTIONS

1. In your opinion, what are the most important topics software engineers should learn in their software engineering education? Did you learn these topics during your studies?
2. What are the most important skills a software engineer should have to succeed in the software industry?
3. Assume that you are a prospective software engineering student. Review the Web for several software engineering programs. Which one would you prefer? Why? Explain the main considerations on which you base your decision.

RELEVANCE FOR SOFTWARE ENGINEERING

The relevance of the education of software engineers to the profession of software engineering is obvious. It derives simply from the fact that those who are educated to become software engineers may have some influence on the way the field is shaped. Since the field of software engineering is relatively young (see Chapter 8, "The History of Software Engineering"), this influence may be even more significant than in the case of more mature domains. Indeed, the citations presented at the beginning of this chapter indicate the close interconnection between academia (where software engineers are educated) and the software industry (where the knowledge gained at the university is applied). Based on the literature that discusses this interconnection, it seems that the relevance of the education of software engineers to the field of software engineering is relatively high, and that gaps between the education of software engineers and the way software engineering is accepted in the industry can and should be bridged.

THE HISTORY OF SOFTWARE ENGINEERING EDUCATION[1]

As described in Chapter 8, the term *software engineering* has been widely used since the late 1960s. However, it was only in the 1990s that software engineering gained acceptance as a professional and academic discipline. This recognition is reflected by the growing number of software engineering degree programs in colleges and universities throughout the world.

It was 1976 when Peter Freeman of the University of California, Irvine, and Anthony I. Wasserman of the University of California, San Francisco, organized the first workshop on software engineering education. This workshop was organized as an answer to the gap in the 1970s between the industry that continued to struggle to build larger and more complex software systems and the educators in the universities who continued to create and teach the new discipline of computer science.

One of the milestones in the development of software engineering education was the establishment of the Software Engineering Institute (SEI) at Carnegie-Mellon University, in Pittsburgh, Pennsylvania, in 1984. The main mission of the institute was to provide leadership in advancing the state of the practice of software engineering to improve the quality of systems that depend on software. However, based on the recognized importance of software engineering education in the preparation of software professionals, the institute's charter also required it to "influence software engineering curricula throughout the education community." In 1987, 1989, and 1991, the SEI published model curricula for university Master of Software Engineering (MSE) programs.

As mentioned in this book several times, software engineering is a broad and diverse discipline. Consequently, all of its content cannot be covered in any curriculum. To cope with this challenge, the SEI designers of the MSE curriculum used a broad view of software engineering, including some topics that are not part of a typical engineering curriculum.

The late 1980s and early 1990s produced a steady growth of MSE programs in U.S. universities. Because the term *software engineering* is still controversial, many of the new programs have different titles. The same phenomenon happened in other places on the globe. Although the structures of university curricula differ greatly from country to country, there are many programs worldwide that can be cited as software engineering programs.

THE EDUCATION OF SOFTWARE ENGINEERS TODAY

Efforts to establish a standard software engineering curriculum have continued until today. One of these attempts is presented in the first draft of the "Computing Curriculum—Software Engineering" created jointly by The Joint Task Force on Computing Curricula of the IEEE Computer Society and the Association for Computing Machinery, published on July 17, 2003 (*http://sites.computer.org/ccse/volume/First-Draft.pdf*). This volume is one out of four created by the IEEE Computer Society/ACM Task Force on the Model Curricula for Computing. The job of this Task Force was to review the 1991 curricula and develop a revised and enhanced version that addresses developments in the computing world in the past decade and will sustain it through the next decade. This effort yielded five volumes: an overview volume that describes general principles and commonalities among all of the specific discipline volumes, and four specific discipline volumes (Computer Science, Computer Engineering, Software Engineering, and Information Systems). The effort involved in this task reflects the diversity of the computing field. Additional information can be found at *www.computer.org/education/cc2001/*.

Eleven principles influenced the creation of the software engineering volume in order to capture the special nature of software engineering that differentiates it from other computing disciplines. From these 11 principles we list the following four principles to which, in our view, this book is closely related:

- Software engineering draws its foundations from a wide variety of disciplines (Principle 2).
- Development of a software engineering curriculum must be sensitive to changes in technology, developments in pedagogy, and the importance of life-long learning (Principle 4).

- Guidance on software engineering curricula must be based on an appropriate definition of software engineering knowledge (Principle 7).
- The Computing Curriculum—Software Engineering volume must include exposure to aspects of professional practice as an integral component of the undergraduate curriculum (Principle 10).

The preceding principles are selected because they highlight the connectivity of software engineering to other areas. Inside the community of software engineering some voice the opinion that the way university courses present computer programs to students does not reflect the actual complexity involved in the process of developing computer programs. As emphasized throughout this book, there are two aspects to that complexity. The first refers to the cognitive complexity of the actual process of developing software systems; the second refers to the social complexity of software development environments, which includes, among other factors, customers and team members. In this respect, Denning [Denning92] says that our teaching is organized as a continuing presentation of important facts, methods, and models that are transferred to the students as a subset of the body of knowledge constituting the discipline. He recommends that software engineering educators should recognize a second kind of knowledge that should be addressed in the software engineering education. This knowledge should include topics such as knowing how to listen, to design, to care, to be organized for new learning, to be a professional, and even to be trustworthy and honest. We believe that this book's adherence to this call is clear.

According to Lethbridge [Lethbridge00], a similar conception with respect to software engineering education is expressed by software practitioners. Lethbridge presents the result of a survey of software practitioners conducted during the summer and autumn of 1998. The survey was designed to check what knowledge is important to the participants and to better understand participant educational and training needs. Among other conclusions, the following two findings are relevant to the message conveyed by this book.

The first finding addresses the topics that software practitioners find most important for their work. According to that survey, the 10 (out of 75) most important topics that every software engineer should presumably know are specific programming languages, data structures, software design and patterns, software architecture, requirements gathering and analysis, human-computer interaction/user interfaces, object-oriented concepts and technology, ethics and professionalism, analysis and design methods, and, finally, presenting to an audience.

The second finding presents a list of topics for which, according to the participants, there is a necessity for improvement in university education. The first seven topics in this list are software design and patterns, requirements gathering and analysis, software architecture, human-computer interaction/user interfaces, object-oriented concepts and technology, ethics and professionalism, and, finally, analysis and design methods. Examination of the nature of these topics reveals that most of them belong in the skills and attitudinal components of the discipline.

Indeed, these two findings illustrate the practitioners' perspective of software engineering as a mixture of scientific, technological, and human-oriented dimensions. This book addresses the third dimension.

TEACHING HUMAN ASPECTS OF SOFTWARE ENGINEERING

We close this chapter with presetting an organizational framework that may be applied in the actual teaching of a "human aspects of software engineering" course.

Table 18.1 presents a two-dimensional course framework [Hazzan03]. One dimension addresses the two human aspects of software engineering discussed in this book: the cognitive aspect and the social aspect. The second dimension refers to each of these aspects on the individual level and on the team level. Each cell presents several topics that can be categorized into it and several activities that are appropriate to be conducted with respect to the topic that the cell addresses. Needless to say, the table exhausts neither all the topics discussed in the course nor all the possible activities that can be conducted in such a course.

TASK

1. Review the different chapters of this book. For each cell in Table 18.1, find at least three topics (or activities) that fit into it.
2. For each cell in Table 18.1, add at least one topic and one activity.

We believe that a single course is not sufficient to include all the aforementioned. "Covering" the topics is not the problem, however. Rather, it is important to increase students' awareness of the central role of the human aspects of software engineering in software development processes, and to equip them with learning skills that will enable them to cope successfully with new (related to software engineering) situations, for which they will need to learn new topics. One of the best ways to achieve these goals is by addressing these issues throughout the entire undergraduate education.

Sometimes time constraints force us to squeeze the "human aspects of software engineering" course into a one-semester, or even smaller, timeframe. In such cases, we recommend the course not be based solely on lectures, but rather integrate interactive and reflective activities. Such an interactive nature of the course can be achieved by reflections, class discussions, case study analysis, and student presen-

TABLE 18.1 Two-Dimensional "Human Aspects of Software Engineering" Course Framework

Aspect Level	Cognitive Aspects	Social Aspects
Individual Level	**Topics:** Program comprehension Reflective practitioner Abstraction **Activities:** Personal reflections on topics such as software development process Construction of ladders of reflection	**Topics:** Code of Ethics of Software Engineering Minorities and gender issues Roles in teamwork **Activities:** Examination of individual motives during the process of software development Examination of ethical dilemmas concerning software development processes
Team Level	**Topics:** Code inspection Learning organization Customer requirements **Activities:** Team discussions, analysis, and reflections on code inspection sessions Reflecting on communal learning processes	**Topics:** Teamwork: structures, problems, benefits, and dilemmas Software development methods **Activities:** Analysis of social factors of software development methods and their influence on teamwork Analysis of connections between the structure of software teams and team social interaction

tations. For lecturers' convenience, the Appendix of this book presents a sample of the course slides for several topics.

REFERENCES AND ADDITIONAL RESOURCES

[Denning92] Denning, Peter J., "Educating a New Engineer," *Communications of the ACM* 35 (12), (December 1992): pp. 82–97.

[El-Kadi99] El-Kadi, Amr, "Stop That Divorce!" *Communication of the ACM* 42 (12), (December 1999): pp. 27–28.

[Glass97] Glass, Robert L., "Revisiting the Industry/Academe Communication Chasm," *Communication of the ACM* 40 (7), (July 1997): pp. 11–13.

[Gudivada03] Gudivada, Venkat N., "The Computing Profession at the Crossroads," *Computer* (May 2003): pp. 90–92.

[Hazzan03] Hazzan, Orit, "Cognitive and Social Aspects of Software Engineering: A Course Framework," *Proceedings of the Eighth Annual Conference on Innovation and Technology in Computer Science Education* (ITiCSE 2003), Thessaloniki, Greece (June 2003): pp. 3–6.

[Lethbridge00] Lethbridge, Timothy C., "Priorities for the Education and Training of Software Engineers, *The Journal of Systems and Software* 53(15), (2000): pp. 53–71.

[Tomayko98] Tomayko, James E., "Forging a Discipline: An Outline History of Software Engineering Education," *Annals of Software Engineering* 6 (1998): pp. 3–18.

[Yeh02] Yeh, Raymond T., Educating Future Software Engineers, *IEEE Transactions on Education* 45, (February 2002): pp. 2–3.

Little, Joyce C.; Granger, Mary J.; Adams, Elizabeth, S.; Jaana, Holvikivi; Lippert, Susan, K.; Walker, H. M.; and Young, Elison, "Integrating Cultural Issues into the CS/IS Curriculum," *Working Group Report—ITiCSE 2000* (2000): pp. 136–154.

ENDNOTES

[1] The section about the history of software engineering education is based on the comprehensive paper [Tomayko98].

19 Additional Information on Resources Used in This Book

This chapter elaborates on some of the references mentioned at the end of each chapter as well offering as new references. This elaboration is intended for readers who want to deepen their understanding in the area. The reading is organized by alphabetical order in three lists: books, articles, and Web sites. For each item, we briefly explain what it is about and its relevance to our book.

When we finished editing this chapter, we realized once again the huge number of topics to which the human aspect of software engineering is connected. Thus, we find the following lists additional methods by which we can inspire the importance of being aware of the multifaceted nature of the discipline of software engineering. It is important to note that these lists are neither recommendations nor criticisms. Accordingly, we do not criticize the resources that we look at. We advise readers to review these brief descriptions and decide on the relevance of the resource.

BOOKS

Anderson, David J., *Agile Management for Software Engineering,* **Prentice Hall, 2004.**

The book explains how agile software development methods produce (better) business results by suggesting how to combat the biggest business complaints about software (late, doesn't deliver as promised, over budget, etc.). It is clear how it is connected to our book. These problems cannot be solved unless emphasis is placed on the human aspect of software engineering. One idea that we find important in this book is the notion that all the agile software development methods share a healthy balance between strict process and individual flexibility that can be achieved in software development processes. For a further analysis of software development methods from the human perspective, see Chapter 2, "Software Development Methods," of our book.

Aspray, William, and Campbell-Kelly, Martin, *Computer—A History of the Information Machine,* **Basic Books, 1996.**

As its name indicates, the book lays out the complex and interesting history of computers, which began at a time when computers were people. This history explores the roots of the software industry's phenomenal development. This book adds to our book the historical perspective to which we dedicate only Chapter 8, "The History of Software Engineering." As with other histories, we find it fascinating that the way it is shaped is so largely influenced by human behavior. As observed throughout our book, we aim to deliver a similar message with respect to software development projects.

Auer, Ken, and Miller, Roy, *Extreme Programming Applied,* **Addison-Wesley, 2002.**

This book explains how to apply the eXtreme Programming discipline. It also provides insight into the motivations and principles behind it. Beyond explaining XP's practices, the book also illustrates how the pieces fit together. By doing so, we suggest, the book refers directly to the human aspect of software engineering since it describes plausible resistance that may be proclaimed toward XP.

Beck, Kent, *Extreme Programming Explained,* **Addison-Wesley, 2000.**

This book is the introduction to eXtreme Programming (XP), one of the most accepted agile software development methods. Since XP is so much about human nature, not surprisingly, human nature is mentioned in many chapters of the book. The book describes what XP is, its main principles, and how it works in practice. XP values and practices are demonstrated. The book highlights that although these principles are not new, their synergy fosters a new software development process. In addition to a description of where XP works, the book addresses where it does not, and offers suggestions for migrating teams and organizations to the XP process.

Beck, Kent, *Test Driven Development: By Example,* **Addison-Wesley, 2003.**

The approach outlined in this book aims at reducing software defects and producing software that is more robust. It is based on writing automated testing before and during coding. It is closely related to the human aspects of software engineering since the test-driven development approach helps programmers overcome the tendency to skip tests when there are time pressures and deadlines approach.

Boehm, Barry W., *Software Engineering Economics,* **Prentice Hall, 1981.**

By addressing the management of software projects, this book is mainly connected to Part IV of our book—but not only there. It also specifically addresses topics such as factors that change software estimates; something we also aim to increase readers' attention to their existence.

Brooks, Frederick P., *Mythical Man-Month,* **Addison-Wesley, 1975 (Second Edition 1995).**

This is considered one of the classic books on the human aspects of software engineering. Evidence for this observation is the fact that in the Twentieth Anniversary Edition of the book, Brooks has identified that most of the basic problems of software development processes and management have not been solved. The book is closely related to our book since both refer to the ways the nature of individuals and groups is expressed in software development environments. This book consists of a collection of essays on software engineering and managing complex projects. The Twentieth Anniversary Edition of the book also contains a reprint of Brooks' 1986 paper "No Silver Bullet" (see the section *Articles*) together with his recent comments on that essay.

Brown, Theodore L., *Making Truth: Metaphor in Science,* **Illinois University Press, 2003.**

This book demonstrates the presence and the power of metaphorical thought and illustrates the deeply metaphorical nature of scientific reasoning and communication. We highlight the metaphor issues since it is about human aspects of software engineering. Indeed, XP includes metaphor as one of its practices.

Campbell-Kelly, Martin, *From Airline Reservations to Sonic the Hedgehog: A History of the Software Industry,* **MIT Press, 2003.**

This book gives a comprehensive overview of the software industry in the United States, which started when companies had to write their own programs for early IBM mainframes. Three sectors are examined: software contracting, corporate software products, and mass-market software products. A final chapter examines the reasons for the success of the U.S. software industry. Reasons such as an early start, market size, and government support for R&D are discussed. This book adds a lot of information to our Chapter 8.

Cerruzi, Paul, *A History of Modern Computing,* **MIT Press, 2000.**

This history book covers modern computing from the development of the first electronic digital computer through the advent of the World Wide Web. It focuses on computing and presents stories of individuals and institutions. Emphasis is placed on those factors that conspired to bring about the decisive shifts in the story. This book adds many details to Chapter 8 of our book.

Deimel, Lionel E., and Naveda, J. Fernando, *Reading Computer Programs: Instructor's Guide and Exercises,* **Software Engineering Institute, Carnegie-Mellon University, 1990.**

This guide is closely related to Chapter 9, "Program Comprehension, Code Inspections, and Refactoring," of our book. Both stress the importance of computer program readability. While we examine this issue mainly from the developers' perspective, this guide provides educators with material with which to teach the topic in general and sample exercises to facilitate the teaching of program reading in particular. Although the programming language used in the report is Ada, most of what is said in the guide is independent of any particular language. Instructors who use our book can use this guide as a supplementary material when they teach the topic of program comprehension.

DeMarco, Tom, and Lister, Timothy, *Peopleware: Productive Projects and Teams,* **Dorset House, 1987.**

This book demonstrates that the major issues of software development are human, not technical. In particular, the book addresses specifically the failure of software development projects. These are the main reasons why we find this book so connected to our book. Similar to our assertion, *Peopleware* finds some of the reasons for these failures in team members. The book includes plenty of anecdotes. Readers may find it possible to connect some of them to topics discussed in our book and to use them as a basis for their case studies analysis.

DeMarco, Thomas, *Slack,* **Broadway Books, 2002.**

This book conveys the message that in today's competitive marketplace, managers work far less effectively than before. Managers overemphasize deadlines and rush employees, sacrificing quality. Clearly, this book is connected to our book because it is about human behavior in general and human aspects of software engineering in particular. The term *slack* refers to the degree of freedom in a company that allows it to change. Among other ways, slack can appear in the way a company treats employees: instead of loading them up with overwork, a company designed with slack allows its people room to breathe and reinvent themselves. Accordingly, the book advocates teamwork, discouragement of competition, and allowing training time.

Drucker, Peter Ferdinand, *Managing for the Future,* **Truman Talley Books, 1992.**

The book presents the author's views on the business challenges of the present and the future. The essays included in the book are arranged under four broad headings: Economics, People, Management, and The Organization. Although it is not directly connected to software development, we believe that it is important for any software engineer to be aware of the messages in this book.

Duarte, Deborah L., and Snyder, Nancy Tennant, *Mastering Virtual Teams,* Second Edition, Jossey-Bass, 2001.

This book discusses the management of teams that are not located in one time, geography, culture, and organization zone. With such a focus, the book is closely related to our discussion of virtual teams in Chapter 3, "Working in Teams," and to our Chapter 6, "International Perspective on Software Engineering." The book outlines suggestions applicable to large and small organizations and presents examples together with their analysis. Based on this analysis, checklists and practical exercises are suggested for the interested readers. Some of the examples can be used as a basis for a case study analysis as well as being analyzed from additional perspectives discussed in our book.

Fowler, Martin (with contributions by Kent Beck, John Brant, William Opdyke, and Don Roberts), *Refactoring: Improving the Design of Existing Code,* Addison-Wesley, 2000.

Focused on refactoring, this book ties tightly to Chapter 9 of our book. It suggests a collection of techniques and a detailed catalog of refactoring operations that aim to improve poorly designed programs that result in applications that are inefficient and hard to maintain and extend. The contribution of refactoring both to the social aspect and to the cognitive aspect of software engineering is significant.

Gates, Bill, with Hemingway, Collins, *Business @ the Speed of Thought: Using a Digital Nervous System,* Penguin, 1999.

This book examines the role of information technology in the life of organizations. It illustrates how integrated information systems can help every organization achieve its goals. It focuses on the way in which information technology is used in knowledge management processes, which directly determine the business success. The book presents examples of companies that have already successfully engineered information networks to manage the different aspects of their business (inventory, sales, and customer relationships) in a better way. Although the book examines management activities with respect to information technology on the organization level, we find it suitable to look at this book together with our book, since the processes presented in the book are closely related to the everyday life of software engineers in the organization.

Hamlet, Dick, and Maybee, Joe, *The Engineering of Software,* Addison-Wesley, 2001.

This book is one of the many available that introduce the essential activities involved in a software engineering project. The book is divided into four parts: Software and Engineering, Requirements and Specification, Design and Coding, and

Software Testing. The ideas presented are treated with a multitude of examples. On several occasions the authors examine the nature of the profession of software engineering and reexamine its connection to other engineering professions and to other kinds of professions such as art. We convey a similar message in different places in our book (Chapter 7, "Different Perspectives on Software Engineering"). In such places, we also question the nature of the discipline of software engineering and emphasize that it may be viewed from different perspectives, each highlighting a facet of the discipline.

Humphrey, Watts, *A Discipline for Software Engineering*, **Addison-Wesley, 1995.**

This book presents a perspective on software process management and lays the foundation for a disciplined approach to software engineering. It describes methods for software development on the individual level in order to gain the benefits of software teamwork. The methods presented develop the skills and habits needed to plan, track, and analyze large, complex software projects. The book includes project exercises to give readers the opportunity to practice software process management as they learn it. The book is well connected to Chapters 2 and 3 in our book. We find this book relevant to our book, since the focus of Humphrey's book is on the individual and attention is put on what developers go through in the process of software development.

Humphrey, Watts, *Team Software Process*, **Addison-Wesley, 2000.**

Software development processes can be viewed from the organizational, team, and individual levels. This book examines the process of software development from the team level, addressing topics such as the formation of software development teams, motivation of software developers, productivity issues, role assignments, communication, conflict solving, and tracking and progress reports. The book provides two project exercises, with prescribed development goals and team roles to let the readers experience the book's messages. This book is perfect to read together with our Chapter 3.

Jacobson, Ivar; Booch, Grady; and Rumbaugh, James, *The Unified Development Process*, **Addison-Wesley, 1999.**

The book introduces a UML-based software development method—The Unified Development Process—whose human aspects are examined in Chapter 2 of this book. The software process, which involves everything from gathering system requirements to analysis, design, implementation, and testing, is presented in detail. UML-based examples illustrate the main ideas, including the connection between UML diagrams with various elements used throughout the software process.

Kernighan, Brian, and Plauger, P. J., *The Elements of Programming Style*, **Second Edition, McGraw-Hill, 1988.**

This book is closely related to Chapter 9 of our book. Although it focuses on programming, it is not a specific language programming book. Rather, it is about how to write good programs in any language. We find it appropriate to associate it with our book since it also addresses the software development activity as a craft, and thus invites the consideration of different aspects beyond the technical.

Kerth, Norman L., *Project Retrospective*, **Dorset House Publications, 2001.**

This book guides facilitators and participants in software organizations through the process of the project retrospective. The aim of these retrospectives is to preserve the lessons learned from the success and failure of previous software projects. There is careful attention to how to allay the fear of such a process and how to establish an air of mutual trust. The book is clearly connected to the human aspects of software engineering in general, and to the reflective practice perspective (see our Chapter 10, "Learning Processes in Software Engineering") in particular. All these processes require one to be sensitive and aware of the multifaceted nature of such situations.

Kuhn, Thomas, *The Structure of Scientific Revolutions*, **University of Chicago Press, 1962.**

In this book, Kuhn outlines his theory that the scientific community resists radical, yet valid, theories because they are necessarily subversive of its basic commitments. The stage in which a shift in professional commitments to shared assumptions takes place is described by Kuhn as a scientific revolution. Since the new theories require the reconstruction of prior assumptions and the reevaluation of prior facts, the process is not simple, is strongly resisted by the established community, and is time consuming. Usually, scientists move from disdain through doubt to acceptance of a new theory. Naturally, social and psychological factors are woven into this process, which is why we find it relevant for our book. In a way, the agile perspective toward software engineering is a new paradigm. According to Kuhn, it is natural that agilism raises resistance among the community of software engineers. If agilism proves to be a better way to produce software, it is suggested that it will follow the next stages described in Kuhn's book. Not surprisingly, it has been already argued that Kuhn's book has an influence far beyond its originally intended audience.

Lakoff, George, and Johnson, Mark, *Metaphors We Live By*, **University of Chicago Press, 1980.**

This book argues that metaphor is integral, not peripheral to language and understanding, and that metaphor is pervasive in everyday life, not just in language but also in thought and action. It is connected to our book since we believe that one can increase one's understanding of his or her environment by listening to how people talk in general and their use of metaphor in particular. Furthermore, metaphor can be used as a means for improving communication between teammates and between customers and software team members.

Margolis, Jane, and Fisher, Allan, *Unlocking the Clubhouse: Women in Computing*, **MIT Press, 2002.**

This book is tightly connected to our Chapter 6 in which we discuss gender issues related to the software industry in general. This book examines the phenomenon according to which only 15 percent to 20 percent of undergraduate computer science majors at leading U.S. computer science departments are female. One of the main arguments of the book is that the under-representation of women among the creators of information technology tools has serious consequences, both for those women whose potential contribution is not expressed and for the society that is shaped by that technology. This assertion is illustrated by several examples. The research- and action-oriented portion of the book takes place at the School of Computer Science at Carnegie-Mellon University. For five years (1995–2000), the authors engaged in an interdisciplinary program of research and action in response to this situation. The research examines male and female students' engagement with computer science. The action component aims at effecting changes in the pedagogy and culture that will encourage broader participation in the computing enterprise. At the end of that period, the entering enrollment of women in the undergraduate Computer Science program at Carnegie-Mellon rose from 8 percent in 1995 to 42 percent in 2000. This book may help those readers who are interested in the topic to deepen their understanding of Chapter 6 of our book.

McConnell, Steve, *Rapid Development*, **Microsoft Press, 1996.**

This book is about effectively scheduling software development and how to get high-pressure software development schedules under control. Attention is devoted to classic mistakes of which developers, team leaders, and managers should be aware. Among other classic problems, the following are discussed: adding people to a late project, abandonment of planning under pressure, planning to catch up later, and the "Silver Bullet" syndrome. In addition to laying out these problems, the book describes concrete solutions and tips. The book also includes illustrative case studies. It is clear how this book is related to ours. Both books deliver the message that some of the well-known problems of software development are people centered, not technology based.

Naur, Peter, and Randell, Brian (eds), *Software Engineering—Report on a Conference Sponsored by the NATO Science Committee,* **Garmisch, Germany, October 7–11, 1968, Brussels, Belgium, 1969.**

This book has historical value. It is the proceedings of the NATO conference that took place in the fall of 1968 (and again in the fall of 1969), devoted to the subject of software engineering, in which the term *software engineering* was coined. These conferences were motivated by the troubles that the computer industry faced when producing large and complex software systems. Not surprisingly, some of the problems are still recognized in current software projects. The relevance of this book to our book is obvious. First, the need for these conferences emerged from people's need, a clear human topic. Second, from the historical angle, it constituted the field of problems we are still struggling with.

Shapiro, Carl, and Varian, Hal R., *Information Rules,* **Harvard Business School Press, 1998.**

This book applies the economics of information and networks to practical business strategies. It offers a perspective of how economic systems work, examining the underlying economic forces that determine success and failure. The book describes the authors' perspective on how to do business in the information age and covers issues such as pricing, intellectual property, versioning, and standards. Although the book is not directly connected to our book, we include it here, because we believe the messages it delivers may help any software engineer's navigation in the software industry.

Schön, Donald A., *The Reflective Practitioner,* **BasicBooks, 1983.**

Schön's book is very relevant for our book. One of the messages in our book is to increase readers' awareness to the potential contribution of the intertwining of a reflective mode of thinking into the software development process. Although this message is spread all over our book, it is highlighted in Chapter 10. In his book, Schön examines five professions—engineering, architecture, management, psychotherapy, and town planning—to illustrate how professionals go about solving problems, how "reflection-in-action" works, and how this skill might be fostered in future professionals. As far as we know, Donald Schön is one of the few people who have written about reflection and its role in professional life. His perspective has a lot to contribute to the profession of software engineering.

Schön, Donald A., *Educating the Reflective Practitioner: Towards a New Design for Teaching and Learning in The Profession,* **Jossey-Bass, 1987.**

This book expands on Schön's previous book, *The Reflective Practitioner.* In the first book, Schön outlines his perspective about the reflective practice perspective; in this book, he focuses on how to educate students to become reflective practitioners. He does so by focusing on several coach-student interactions in which the coach aims at elevating the student's reflective skill. Since we believe that reflective skills may improve the performance of software engineers, we find this book extremely relevant for our book. We especially recommend those readers who teach a Human Aspect of Software Engineering course to read this book. We hope that this perspective will trickle into their course as well as send a message to prospective software engineers.

Senge, Peter M., *The Fifth Discipline: The Art and Practice of the Learning Organization*, Currency Doubleday, 1990.

This book about learning organization is tightly connected to Chapter 10 of our book. In particular, we emphasize the learning organization perspective with respect to software organizations. The idea of the book is that organizations use the systems thinking method to become a learning organization. Analysis of the spirit of the profession of software engineering suggests very clearly that software organizations should acquire some of the principles of learning organizations.

Senge, Peter M., *The Fifth Discipline: Fieldbook*, A Currency book, published by Doubleday, 1994.

This book applies the principle of Senge's previous book by guiding existing companies on how to become learning organizations. The book presents exercises for both individuals and teams, suggests approaches and ideas, and tells success stories. All can be integrated into software organizations, of course, in order to start to overcome the basic problems that characterize software projects.

Shneiderman, Ben, *Software Psychology—Human Factors in Computer and Information Systems*, Winthrop Publishers, Inc., 1980.

As far as we know, this book is the first to introduce the psychological perspective into computer science and software engineering. It addresses topics such as programming as human performance, programming style, software quality evaluation, group processes, and designing interactive systems. It is clear how this book relates to our book. The difference, however, lies in our emphasis on the actual process of software development as it is experienced by software engineers, while Shneiderman's book aims to introduce psychological methods into the practice of software development. Although the focus is somehow different, we see the two books as complementary.

Williams, Laurie, and Kessler, Robert, *Pair Programming Illuminated*, Addison-Wesley, 2002.

Pair programming is one of the XP practices. It invites many debates since many mangers argue that it is not an effective way to produce software. This practice has many benefits, however. Since two programmers continuously collaborate on the same design, algorithm, code, and test, the code they produce is of higher quality. The resistance that this practice raises connects it very tightly to the human aspect of software engineering. Here are two relevant questions. First, why does it raise resistance? Second, how can one convince software engineers to try this practice before they reject it? This book explains both the principles underlying this method and its best practices, and illuminates the main ideas with two case studies. Again, as with other XP practices, we observe that while some practices on the surface look like a waste of effort, their deeper analysis reveals their potential contribution to software development processes.

Winograd, Terry (ed.), *Bringing Design to Software*, Addison-Wesley, 1996.

This book shows how to improve the practice of software design by applying lessons from other areas of design to the creation of software. These lessons aim at creating appropriate and effective software that works. The book consists of essays contributed by software and design professionals, interviews with experts, and profiles of successful projects and products. All these together illuminate what design is and indicate what the core of all the design practices is. We connect this book to our discussion in Chapter 10 with respect to a reflective practice perspective. However, we believe that it can contribute to the discussion about the human aspects of software engineering from additional perspectives.

Yourdon, Edward, *Decline and Fall of the American Programmer*, Prentice Hall, 1992.

This book and Yourdon's *Rise and Resurrection of the American Programmer* explore conclusions with respect to the direction of software engineering and the career prospects of the American programmers and the production of software in other places on the globe. We mention this book in Chapter 6 where we explore connections between different places on the globe with respect to software development. We believe that software engineers who care about human aspects of software engineering should increase their awareness with respect to this international perspective on software engineering. As with other topics discussed in our book, we believe that the more one increases one's awareness toward such topics and comprehends that there are many ways to approach them, the better one can cope with the complex situations that are so predominant in software engineering.

ARTICLES

This list is relatively short. We chose to elaborate only on the following articles since we find them directly connected to the global orientation of our book. Many other resources were extremely useful in the process of writing this book. These articles are presented in the relevant chapters where they have been used.

Berry, Daniel M., "The inevitable pain of software development: Why there is no silver bullet," *International Workshop on Time-Constrained Requirements Engineering,* **2002.**

This paper refers directly to Brooks' famous paper "No Silver Bullet: Essence and Accidents of Software Engineering." The paper reviews a variety of programming models, methods, artifacts, and tools and examines them to determine that each has a step that programmers find painful enough that they usually avoid or postpone the step. The paper concludes that, hence, there is no silver bullet.

Brooks, Fredrick P., "No Silver Bullet: Essence and Accidents of Software Engineering," *IEEE Computer* **(April 1987): pp. 10–19.**

This article argues that no silver bullet exists for software project management by examining both the nature of the software problem and the properties of the bullets proposed. In a similar spirit to our book, the discussion in Brooks' article addresses the fact that software is very different from tangible goods.

Camp, Tracy, "The Incredible Shrinking Pipeline," *Communications of the ACM* **40(10), (October 1997): pp. 103–110.**

This paper addresses specifically the pipeline that represents the ratio of women involved in computer science from high school to graduate school and then at the bachelor's level. The importance of the topic is emphasized by the fact that the number of women at the bachelor's level affects the number of women at levels higher in the pipeline and in the job market. The paper also speculates on what the future holds.

Fairley, Richard E., and Willshire, Mary Jane, "Why the Vasa Sank: 10 Problems and Some Antidotes for Software Projects," *IEEE Software* **(April–May 2003): pp. 18–25.**

This article examines the reasons for the warship Vasa's sinking on August 10, 1628. This event is explained by problems in its design and construction process that are remarkably relevant to modern-day attempts to build large, complex software systems. The article describes the problems encountered in the Vasa project, and interprets the problems encountered in terms of modern software projects.

Lethbridge, Timothy C., "Priorities for the Education and Training of Software Engineers," *The Journal of Systems and Software* **53(15), (2000): pp. 53–71.**

This article presents the results of a survey of almost 200 software developers who were asked about which types of knowledge they found most useful in their work, and what they learned in university. The responses are analyzed from different perspectives to better understand the variability needed in the education and training of software engineers.

WEB SITES

These Web sites are selected from those presented at the end of the different chapters of the book. We focus here on Web sites that may serve as a portal for the relevant topic.

Software Engineering Institute: *www.sei.cmu.edu*

The Software Engineering Institute (SEI) is a research and development center sponsored by the U.S. Department of Defense. It aims to provide the technical leadership to advance the practice of software engineering. One of its main activities was the formulation of The Capability Maturity Model (CMM) for software (*www.sei.cmu.edu/cmm/cmm.html*) for judging the maturity of the software processes of an organization. The CMM model helps organizations identify the key practices required to help them increase the maturity of their software development processes.

The Software Engineering Ethics Research Institute: *http://seeri.etsu.edu/*

This Web site provides many resources to increase awareness of the development of ethical and professional practices that address the impacts of software engineering and related technologies on society.

The AgileAlliance: *www.agilealliance.com/home*

The AgileAlliance is a nonprofit organization dedicated to promoting the concepts of agile software development, and helping organizations adopt those concepts. Its main principles are outlined by the Manifesto for Agile Software Development (*www.agilemanifesto.org/*). This community is important for those who are interested in agile software development methods (see Chapter 2).

Refactoring Home Page: *www.refactoring.com*

This site is a portal for information on refactoring. It contains general information about refactoring, a catalog of common refactorings, and tools to help the refactoring process.

ACM's Committee on Women and Computing: *www.acm.org/women/*

This committee of the ACM on Women in Computing informs and supports women in computing. It has contacts with computer scientists, educators, employers, and policymakers to improve working and learning environments for women. The Web site presents resources about the topic and relevant information.

The Association for Women in Computing (AWC): *www.awc-hq.org/*

The Association for Women in Computing is a nonprofit professional organization for women and men who have an interest in information and technology. The Association is dedicated to the advancement of women in the technology fields.

The Knowledge Management Resource Center: *www.kmresource.com/*

This site offers a comprehensive collection of knowledge management resources. In The Knowledge Management Explorer, the material is organized in 17 categories; in the Knowledge Management Bookstore, 219 titles about knowledge management, organizational intelligence, and closely aligned areas of thought and practice are listed.

Appendix: Course Slides

This Appendix presents a sample of the course slides for the following topics:

- Introduction
- The nature of software development projects
- Teamwork in software engineering
- Code of ethics of software engineering
- Software as a product
- Different perspectives of software engineering
- Learning perspectives of software engineering

The entire PowerPoint presentation can be downloaded from *www.charlesriver.com/*. Just search for the book title or authors.

COMMENTS:

- When appropriate, it is recommended to integrate discussions into the course. Some examples are presented in the following slides.
- Naturally, not all slides for each of the above topics are presented. Sometimes, we left place holders that should be completed by the lecturers according to their preferences and their students' background.
- Font size is reduced for the presentation in the book.
- It is recommended to start each lesson with a layout of the course, referring to the exact place in the course flow.

INTRODUCTION TO HUMAN ASPECTS OF SOFTWARE ENGINEERING (CH1)

Slide 1

> # Human Aspects of Software Engineering
>
> Jim Tomayko   Orit Hazzan

Slide 2

> ### What is this course about?
>
> The course highlights the world of software development from the perspective of the people involved in software development processes
> - the individual
> - the team
> - the organization
> - the customer

Slide 3

> ### What is this course about?
>
> Suggest three human oriented topics related to each of the following actors
> - the individual
> - the team
> - the organization
> - the customer

Slide 4

> ### What is this course about?
>
> - Topics:
>
>   [To be completed by the lecturer according to his/her preferences]

Slide 5

> ### How the Course is Taught?
>
> - Reading
> - Short lectures
> - Class activities
> - Class discussions
> - Reflections
> - 2 Homework assignments
> - Case study analysis and presentation
>
> [Grading policy should be added by the lecturer]

Slide 6

> ### Main Message
>
> What?
> - Awareness
> - Reflection
>
> Why?
> - Software complexity
> - Cognitive complexity
> - Social Complexity

Slide 7

The Nature of Software Engineering

- Approaches to SE
- CMU / SEI
- SWEBOK
- Computing Curricula, Software Engineering, CCSE

All reflect the complexity of software engineering

[Lecturers may elaborate on each of these approaches]

Slide 8

The Nature of Software Engineering

SE is a combination of engineering, scientific and social thinking.

- The engineering aspect is important when we examine software systems as products and as tools.
- The scientific aspect is reflected in the problem-solving aspect of software engineering.
- The social aspect is significant with respect to management issues and communication problems.

Slide 9

Two-Dimensional Course Framework

Aspect Level	Cognitive Aspect	Social Aspect
Individual Level		
Team Level		

THE NATURE OF SOFTWARE DEVELOPMENT PROJECTS (CH2)

Slide 1

The Nature of Software Development Projects

Jim Tomayko   Orit Hazzan

Slide 2

Failure & Success of Software Projects

Is the following project a success or a failure?

- Software for a medical equipment
- 64% of the developers had more than 5 years of experience in software development
- Time to market: 193% (of what was expected)
- Over budget: 200%
- After delivery: The project works as was expected

Slide 3

Failure & Success of Software Projects

Is the following project a success or a failure?
- Software for airlines management
- 64% of the developers had more than 5 years of software development experience
- Delivered on-time
- No "over budget"
- After delivery: The project did not work as was expected

Slide 4

Failure & Success of Software Projects

Discussion:
- Define a successful software project.
- Define a failed software project.
- How can the success of a software project be measured?

Slide 5

Why Do Software Projects Fail?

- Problems in teamwork?
- Technical problems?
- Other problems?

Slide 6

Why Do Software Projects Fail?

Schedule
- Stressed time table
- No defined deadlines
- Attempts to overcome this stress
  Shortcuts (skip tests)
  Long hours

Redundant extensions
- tendency to add features (beyond requirements);
- these extensions are not considered when the project time table is set

Slide 7

Why Do Software Projects Fail?

Teamwork communication and cooperation
- Cooperation among team members
  —Job security; information hiding; complex code
- Problems in leadership
- Low quality
  —Sloppy understanding of the task & the requirements

Slide 8

Why Do Software Projects Fail?

Fairley, R. E. and Willshire N. J. (2003). "Why the Vasa sank: 10 problems and some antidotes for software projects," *IEEE Software*, pp. 18–25.

[Lecturer can ask students to review these problems]

The course focuses on different ways for solving some of these problems.

TEAMWORK IN SOFTWARE ENGINEERING (CH3)

Slide 1

Teamwork in Software Engineering

Jim Tomayko   Orit Hazzan

Slide 2

Main Problems in Teamwork

- Suggest general problems of teamwork.
- Suggest problems of software teams.
- Suggest solutions to the problems you mentioned.

Slide 3

What Does It Mean to Be a Team Member?

- Different from working alone

   Instead of being efficient as much as possible, the target is that the team would be efficient as much as possible.

   A gelled team is capable of performing tasks that can not be carried out by the individual(s)
- Shared goals

Slide 4

The Mythical Man-Month (Brooks)

One of the main problems of software development is that schedule is based on Person Months.

Correct for other professions where
- Where a task can be shared among many people
- No communication is needed
- No dependency among people/tasks

Incorrect for software development
- In software development people and months are not interchangeable

Slide 5

The Mythical Man-Month

Brook's law:

Adding manpower to a late software project makes it later.

Slide 6

Communication

Inside and outside the team

- Feedback
- Visibility
- Listening
- Negotiating

Slide 7

Types of Software Teams

Source: Armour, P. G. (2001). "Matching process to types of teams," *CACM* 44(7), pp. 21–23.

Four main types of software teams:

- Technical (its job is to do something)
- Problem-solving (its job is to fix something)
- Creative (its job is to build something)
- Learning (its job is to learn something)

Slide 8

Types of Software Teams

Usually software teams are more than one type:

Example: a team is a
- Learning team when data about the customer is collected;
- Creative team when the system is designed;
- Technical team when they code;
- And a Problem Solving team during debugging sessions.

Slide 9

Structures of Software Teams

- Structure of software teams depends on the type of project and people.
- Sometimes the structure of the project is reflected into the structure of the team
- 3 typical structures:
    - Democratic Team
    - Chief Programmer Team
    - Hierarchical Team

In each team: 3–8 people; enables communication

[Lecturers can elaborate on each team structure according to students' background]

Slide 10

Additional Topics Related to Team Work

Team members heterogeneity:

- Personal variables: gender, age
- Skills
- Previous experience

Team evaluation
Decision making processes
Dilemmas in teamwork (in the continuation)

Slide 11

Dilemmas in Teamwork–Discussions

One of the team members does not work on his/her task.

- What are you doing?

Additional dilemmas from your experience

Slide 12

Relationship Between Reward and Cooperation

Optimum ways for sharing bonuses:
- Shared—as a team member
- Individually—depending on personal contribution

[Lecturers may proceed with the reward allocation activity descried in Chapter 3]

SOFTWARE AS A PRODUCT (CH4)

Slide 1

Software as a Product

Jim Tomayko   Orit Hazzan

Slide 2

- Why do we write software?
- What are the targets of software development?

Discussion Framework

We examine software that is written for customers.

The target: Supply the customers the software they need!

Similar to the case of house construction.

(Look at: *Worldwide Institute of Software Architects*)

Slide 3

Discussion Framework

One of the course's aims: Examination of different points of view at software development processes.

Two topics:
- Requirements management
- Data collection tools

Slide 4

Requirements Management

- A developing area
- Efficient requirements management improves also the communication between the development and marketing departments
- Sometimes, the marketing department promises customers features that the development department is not aware of

Slide 5

Requirements Management

- Requirements management is a strategic decision of a company
- In many cases: If requirements are managed properly, all the development continues smoothly
  —The case of eXtreme Programming

Slide 6

Requirements Management Tools

- Web-based tools, Data bases.
- Characteristics: From Chapter 4.
  —Accessibility
  —Tracing the requirements and their development
- Knowledge is Power—How is it connected to our course?
- The introduction of requirements management tools sometimes raises resistance as those who have the knowledge, loss of their power.

Slide 7

Data Gathering Tools

- Documents
- Interviews
- Observations
- Questionnaires

—In many cases more than one tool is used.
—The literature about the topic is vast.
—Here we see some general guidelines.

Slide 8

Data Gathering: Documents

Become familiar with the organization:
- Structure
- Roles
- Procedures, etc.
  —May be helpful, for example, for selecting those who will be interviewed.

Slide 9

Data Gathering: Interviews

- Important tool for requirement gathering
- Interviews give us:
  - Information
  - Estimations
  - Opinions
  - Feelings

Sometimes people beautify the reality.
Hence, different data gathering tools should be used.

Slide 10

Data Gathering: Interviews

Relevant issues:

- Selecting the interviewees
- Interviewer-interviewee relationships
- Order of interviews
- Kinds of questions (open/close)
- Kind of interview (structured, semi-structured)

Slide 11

Data Gathering: Questionnaires

Questioners are effective when:

- There is a need to collect information from many people
- People are located in different places (and it is impossible to interview all of them).

Important: Pilot!

When the rate of answers is low! Be careful!

Slide 12

Data Gathering: Observations

Fits to situation/professions such as services

Observation teaches us about:
- The kind of work
- How long the work takes
- Relationships between workers
- Employee-customer relationship.

Slide 13

Data Gathering: Observations

Location (fixed/not fixed)

- Structured/not structured
- Problem: People may change their behavior.

Slide 14

Class Activity

Select a company for which you are going to develop a web-based system.

Develop a set of data collection tools in order to collect data about the company.

CODE OF ETHICS OF SOFTWARE ENGINEERING (CH5)

Slide 1

Code of Ethics of Software Engineering

Jim Tomayko   Orit Hazzan

Slide 2

Ethics and Software Engineering

- What is ethics?
- Why is ethics needed?
- Does the Software Engineering community need ethics?

  If yes: On what principles should the ethics of software engineers be based on?

  If not: Explain why

Slide 3

What is Ethics?

The *Webster's Collegiate Dictionary*:

Ethics is "the discipline dealing with what is good and bad and with moral duty and obligation".

Differences between Ethics and Law.

Slide 4

What is a Profession?

Kasher, A. (2002). "Professional Autonomy and Its Limits, "the *Research in Ethics and Engineering* conference, 25–27 April 2002, Delft University of Technology.

Five basic layers:
- systematic knowledge
- problem solving proficiency
- constant improvement of that knowledge and proficiency
- local understanding of professional claims and methods
- global understanding of professional activity, which is **ethics**

Slide 5

What is Ethics? (Asa Kasher)

Ethics reflects what is appropriate in a professional community

The main resources for any ethics:
- the professional dignity
- the conception of the professional community of the profession's essence
- the requirements of the surrounding social environments (e.g., democratic values)

Slide 6

Does the Software Engineering Community Need Ethics?

The importance of ethics of software engineers stems from the significant influence of computers and software on the world.
- On the one hand, software engineers can cause damage;
- On the other case they can contribute a lot to the society.

Software engineers should know what is permitted, what is forbidden, when they have the freedom to choose, etc.

Slide 7

Ethics of Software Engineering

The importance of Ethics in the start-up era:

Martin, C. D. (2001). "Ethics@Coms: Why Internet start-ups need ethics codes," *SIGCSE Bulletin*, 33(2), pp. 7–8.
- Focus was placed on the product and marketing.
- Less attention was put on topics such as vision, mission, ethics.

Ethics contributes to the image of the company; lack of ethics may cause damage.

Slide 8

Ethics of Software Engineering

In what follows, several cases related to software engineering are presented. With respect to each scenario:
- Express your opinion about the described behavior—Is it ethical?
- Describe how would you behave in such a case.
- According to your decision: Formulate one or more ethical norms, which, in your opinion, should be included in the Code of Ethics of Software Engineering.

[Lecturers are invited to select case studies for discussion from the book as well as from their own experience]

Slide 9

Ethics of Software Engineering Class Activity

Formulate ethical roles for the profession of software engineering.

[After that class activity, the code of ethics should be presented to the students. You can use the URL *http://www.acm.org/serving/se/code.htm*]

Slide 10

Comments on the Code of Ethics of Software Engineering

- In the previous version, the first principle was the product.
- In the updated version, the first principle is public.
- The Code is phrased in a way that it fits to changes in the world of software engineering

Slide 11

> **Ethics of Software Engineering–Group Work**
>
> Suggest three activities that can be done on a regular basis to fulfill the Self section of the Code of Ethics of Software Engineering.

DIFFERENT PERSPECTIVES ON SOFTWARE ENGINEERING (CH6 & CH7)

Slide 1

> **Different Perspectives on Software Engineering**
>
> Jim Tomayko   Orit Hazzan

Slide 2

> **What Is This Lesson About?**
>
> Frameworks:
> - Historical perspective
> - Process vs. product perspectives
> - International perspective (Gender)

Slide 3

> **Software Engineering**
>
> The term software was has first used in the 1968 North Atlantic Treaty Organization (NATO) software conference in Garmisch, Germany.
>
> Following the recognition of the software crisis: Recognition that software development is a long and complex process.

Slide 4

> **Software Engineering**
>
> Until today:
> - The Evolutionary Development Model
> - The Iterative Development Model
> - The Spiral Software Development Model
> - Rational Unified Process
> - Extreme Programming
>
> What is common to all these methods?

Slide 5

Software Engineering

All traverse the activities of:

    specifying, designing, coding, testing. Some do this several times.

- These activities have become the **paradigm** of software development.
- Paradigm is *the* way of doing things

Kuhn, T. S. (1962). *The Structure of Scientific Revolutions,* Chicago: Univ. Chicago Press.

Slide 6

Software Engineering

The four activities: specifying, designing, coding, and testing; plus the concepts of abstraction and information hiding served as the basis for iterative methods

When the majority of developers accepted this, it became a paradigm

Slide 7

International Perspective on Software Engineering

Main messages of the lesson:

- Local events may have global influence on the hi-tech sector (Sept 11, SARS)
- Countries may use the software industry as a leverage for their economy
- Gender and minorities in the software world **(the focus of the lesson)**

Slide 8

International Perspective of Software Engineering

- Women and minorities are underrepresented in the community of the IT developers
- We focus on women in the software industry:
  - Women comprise about 50% of the world population, half of the workforce, only 20% of the IT sector
  - The absence of women in the hi-tech industry is almost a (western) worldwide phenomenon
  - Each country has its own minorities; local factors influence their involvement in the IT economy

Slide 9

Women in the Software Industry–Background

- 1983–1994: USA, the shrinking pipeline

Tracy Camp (1997). "The Incredible Shrinking Pipeline," *Communications of the ACM* 40(10), pp. 103–110

- The phenomenon is also identified during 1993–2002:

Vanessa Davies and Tracy Camp (2000). "Where Have Women Gone and Will They Be Returning," *The Computer Professionals for Social Responsibility Newsletter.*

Slide 10

Women in the Software Industry

Actions in order to increase the % of women who study computer science

- CMU, 1995: An interdisciplinary program of research and action has been initiated
- The research aspect aimed to understand students' attitudes towards computer science
- The goal of the action component was to devise and effect changes in curriculum, pedagogy and culture that will encourage the participation of women in the computing world

Slide 11

Women in the Software Industry

Results:
- The entering enrollment of women in the undergraduate computer science program at Carnegie Mellon University has been raised from 8% in 1995 to 42% in 2000
- The full story is described in Margolis, J. and Fisher, A. (2002). *Unlocking the Clubhouse: Women in Computing,* MIT Press, 2002

Slide 12

Women in the Software Industry

Class discussion:
- How would you explain the fact that the percentage of women who study computer science and software engineering is relatively low? What factors may influence their choice of a profession?
- How would you approach the problems?

Slide 13

Women in the Software Industry

Why women do not study CS and SE?
- The image of these professions:
  - Only "nerds" work in the area
  - A typical workday is made of long hours of coding, coding and coding, without any human interaction.
  - This course shows that this image is incorrect.
- Women tend to prefer jobs that are based on human interaction.

Slide 14

Women in the Software Industry

Why women do not study CS and SE? (cont)
- Early education of girls:
  - Girls are educated to accept the traditional, not to take risks and not to compete.
- To succeed in the hi-tech market one has to be an inventor, to be a risk-taker and to compete.
- Young girls' exposure to technology.
- Girls are not encouraged to play with computers.
- Compare the variety of the computer games that are offered to girls versus to boys!

Slide 15

Women in the Software Industry

Food for thought/Discussion:
- Outline main features of a computer game that in your opinion fit and may appeal to young girls.
- Conduct the same task for a game that fits and may appeal to young boys.
- Are there differences between the two games? If yes—what are the differences? If not—what is common to these two games?

Slide 16

Women in the Software Industry

Why women do not study CS and SE? (cont)
- Lack of role models—women with whom the young women can identify.
- Role models may increase the young girls' attention to women's success in the hi-tech area.
- Programs of pairing up female high-school students with mentors in the industry aim to let the young girls experience what a career in the tech fields means.

LEARNING PERSPECTIVES ON SOFTWARE ENGINEERING (CH10)

Slide 1

Learning Perspectives on Software Engineering

Jim Tomayko   Orit Hazzan

Slide 2

Learning Perspectives on Software Engineering

- Reflective practices
- Program comprehension
- Additional important topic: Learning organization in general and learning software organizations in particular.

Slide 3

Software Engineering as a Reflective Practice

Main Idea
- On-going reflection on what we do in the process of software system developing
- Reflection:
  - reflection-in-action: During the development process
  - reflection-on-action: After the completion of the development

Slide 4

Reflective Practice

Schön, D. A. (1983). *The Reflective Practitioner*, BasicBooks.
Schön, D. (1987). *Educating the Reflective Practitioner: Towards a New Design for Teaching and Learning in The Profession*, SF: Jossey-Bass.

Schön wrote about architecture
- Schön did not talk about SE (though he published his books when the software crisis was known)
- Application to disciplines such as management, psychoanalysis
- Two applications to software engineering:
  - The studio method of teaching
  - On-going reflection

Slide 5

Analogy: Architecture Design and Software Design Processes

Worldwide Institute of Software Architects
(*http://www.wwisa.org/wwisamain/role.htm*):
- Phases which define the role of the architect in the software construction process, conceptually following the phases of building construction and architectural services
- Do not fit completely the agile paradigm, yet convey the importance of communication and customer input during software development processes

Slide 6

The Reflective Practitioner Framework and its Relevance to Software Engineering

[Instructors may select several quotes from Schön's book to illustrate the possible application of a reflective practice perspective to software engineering]

Slide 7

Program Comprehension
Introductory Questions

- How and why is the topic connected to our course?
- What does it mean to understand a computer program?
- What factors influence program comprehension?
- In what cases is it important to understand a computer program?
- Why the topic should be discussed at all? Is code reading different from reading a regular text?

Slide 8

Program Comprehension

What does it mean to understand a computer program?

- Knowing all its variables?
- Knowing all its procedures? Functions? Objects? Methods? Data structures?
- Knowing how the program behaves?

Formulate your definition for program comprehension

Slide 9

Program Comprehension

Fjeldstad, R. K. and Hamlen, W. T. (1983):

In making an enhancement or maintenance programmers studied the original program:

- About 3.5 times as long as they studied the documentation of the program, and
- Just as long as they spent implementing the enhancement.

Slide 10

Program Comprehension

Oman, P. W. (1990): Four experiments demonstrate the influence of typographic style (source code formatting and commenting) on program comprehension.

- About one half of the maintenance programmer's time is spent studying the source code and related documentation.

Slide 11

Program Comprehension

Factors affecting program readability
- Reader characteristics
- Representational factors: Programming language, the nature and inclusion of comments
- Typographic factors: Upper and lowercase, fonts, white space
- Environmental factors: The medium of the program (monitor, paper), external documentation, software tools.
- Logical complexity

Slide 12

Program Comprehension

Brooks, R. (1983): Four sources of variation in behavior on program comprehension:
- The kind of computation the program performs
- The intrinsic properties of the program text (length, etc)
- The reason for which the task is performed (modification, debugging . . .)
- Differences among the individuals performing the task

Slide 13

Program Comprehension

Littman, Pinto, Letovsky & Soloway (1987):
Two strategies for program understanding:
- As-needed strategy: Focus on local program behavior. Since the programmer does not approach the modification task with a good understanding of the program, it becomes necessary to gather additional information while the program modification is performed.
- The systematic strategy: The programmer learns how the program behaves before attempting to modify it. Such knowledge permits programmers to design modifications that take these interactions into account. This strategy leads to successful program modification.

Slide 14

Program Comprehension

Vans, A. M., von Mayrhauser, A. and Somlo, G. (1999).
von Mayrhauser, A. and Vans, A. M. (1993).

- Three levels of abstraction with respect to program comprehension (Actions during maintenance):
- Program Level (code)
- Situation level (algorithmic)
- Domain level (application)

Full lists of actions for each level can be found in the book

Slide 15

Program Comprehension– Findings (Cont)

- Switches during corrective maintenance occur slightly more between program and situation models—the two lower levels of abstraction
- Programmers with little experience in the domain, work at lower levels of abstraction until enough domain experience allows them to make connections from the code to higher levels of abstraction

Slide 16

Program Comprehension– Findings

- Programmers with domain experience but little knowledge about the software, also work at lower levels of abstraction, but use their knowledge of the domain to make direct connections into the program.
- Code size affects the level of abstraction on which maintenance engineers prefer to concentrate while building a mental representation of the program. As the code size increases subjects work less with low level program details than higher level functional descriptions of code.

Slide 17

Program Comprehension– Discussion

- How and why is the topic connected to our course?
- Why the topic should be discussed at all? Is code reading different from reading a regular text?

Index

A

abstraction
    about, 129–30
    codes of ethics and, 191
    human aspects of software engineering and, 187–88
    international perspectives and, 191
    learning processes and, 192
    popularity of, 134–35
    program comprehension and, 192
    programming style and, 209
    software architecture and, 193–97
    software development methods and, 188–89
    software requirements and, 190
    software teams and, 189–90
    teaching, 197–98
ACM's Committee on Women and Computing, 316
Ada, 134–35
AgileAlliance, 315
Agile Management for Software Engineering (Anderson), 303
Agile Manifesto, 196
agile software development. *See also* eXtreme Programming
    democratic teams and, 33
    requirements elicitation in, 236–38
    software architecture and, 194–97
    vs. heavyweight approach, 119–20
analog computers, 126–27
Analytical Engine, 134
Anderson, David, 303
Apple Computer, 246
Application Composition model, 233
Architectural Tradeoff Analysis Method (ATAM), 196
Aspray, William, 304
Association for Women in Computing (AWC), 316
Auer, Ken, 304

B

Beck, Kent, 121, 304
Berry, Daniel, 314
Boehm, Barry, 15, 136, 304–5
Booch, Grady, 138, 308

Borg, Anita, 110
bottom-up programming, 184
Bozo Effect, 50–51
Bringing Desire to Software (Winograd), 313
Brooks, Frederick, 4, 14, 116, 228, 245, 305, 314
Brooks' Law, 18
Brown, Theodore, 305
Business @ the Speed of Thought (Gates and Hemingway), 307
business cases, 247–48
business plans, 248–49

C

CAD (computer-Aided Design), 194
Camp, Tracy, 108, 314
Campbell-Kelly, Martin, 304, 305
Capability Maturity Model (CMM), 49–50
Carnegie Mellon University, 108, 109
CASE (Computer Aided Software Engineering), 4
case studies, 269–81
    based on field studies, 292–93
    based on past events, 293–94
    construction of, 286–94
    general principles, 277–80
    presentation of, 294
    project management, 271–75
    relevance of, 284–85
    software development, 274–77
    theoretical, 286–92
    uses of, 285–86
CBAM (Cost/Benefit Analysis Method), 196
CCB (Configuration Control Board), 24
CCSE (Computing Curricula Software Engineering), xxiii, 198, 298
Cerruzi, Paul, 305
Chief Programmer teams, 37
Clark's Method, 230
CMM (Capability Maturity Model), 49–50
COCOMO II, 230–33, 231
code
    See also programming styles
    characteristics of, 203–7

333

debugging, 211–18, 279, 280
    readability of, 280
code inspections (code review), 144–45, 151–54, 166–67
Code of Ethics of Software Engineering, 73–96
    abstraction and, 191
    Code Signature, 94
    full version of, 79, 81–94
    Principle 1, Public, 83–84
    Principle 2, Client and Employer, 84–85
    Principle 3, Product, 85–87
    Principle 4, Judgment, 87–88
    Principle 5, Management, 89–90
    Principle 6, Profession, 90–91
    Principle 7, Colleagues, 92
    Principle 8, Self, 93
    relevance of, 75–77
    scenario analysis and, 77–79
    short version of, 79–80
coding standards, 21
comments, 207–8
communication
    during development process, 20–21
    Internet, 261–62
competition, review of, 248–49
Computer—A History of the Information Machine (Aspray and Campbell-Kelly), 304
Computer-Aided Design (CAD), 194
Computer Aided Software Engineering (CASE), 4
computer industry, 243–50
    See also high-tech industry
computer science
    education, 197–98
    women in, 107–10
computer viruses, 262
computing, early days of, 126–29
Computing Curriculum-Software Engineering (CCSE), xxii, 198, 298
Configuration Control Board (CCB), 24
Cost/Benefit Analysis Method (CBAM), 196
courage, 21–22
customers
    development methods and, 133
    gathering information about, 59–66
    on-site, 19, 21
    Quality Attribute Workshop (QAW) and, 195
    requirements of, 55–70

D
data collection tools
    interviews, 59–63
    observation, 65–66
    questionnaires, 63–65
Davies, Vanessa, 108

debugging, 211–18, 279, 280
Decline and Fall of the American Programmer (Yourdan), 101, 313
Deimel, Lionel, 306
Delay Line, 128
DeMarco, Tom, 306
democratic teams, 31–34
Development Managers, 238
Discipline for Software Engineering, A (Humphrey), 308
distance education (DE), 261–62
documentation, 196–97
dot-com boom, 246
dot-com crisis, 102–3, 227
Druker, Peter Ferdinand, 306
Duarte, Deborah, 307

E
earned value estimates, 234
earned value tracking, 9–10
Eckert, J. Presper, 127–28
e-commerce, 251–59
    about, 254–56
    cognitive analysis of, 256–58
    hypertext and, 257–58
    metaphors in, 256–57
    social perspective of, 258–59
e-communication, 258–59
Educating the Reflective Practitioner (Schön), 164, 311–12
Electronic Delay Storage Automatic Computer (EDSAC), 128, 129
electronic discussion groups, 175
Electronic Numeric Integrator and Calculator (ENIAC), 127–28
Elements of Programming Style (Kernighan and Plauger), 309
Engineering of Software, The (Hamlet and Maybee), 307–8
ethics. *See* Code of Ethics of Software Engineering
Evolutionary Development Model, 136
eXtreme Programming (XP)
    agile software development and, 138–39
    concept of, 18–20
    criteria for choosing, 22–24
    customer requirements and, 57–58
    human aspects of, 20–22
    Planning Game, 235–36
    software architecture and, 194–97
    time-to-market and, 247
Extreme Programming Applied (Auer and Miller), 304
Extreme Programming Explained (Beck), 304

F
Fairley, Richard, 314
feasibility studies, 133

feedback, 21
Fifth Discipline (Senge), 312
Fisher, Allan, 310
Floyd, Christiane, 117
focus groups, 248
Fowler, Martin, 119, 307
Freeman, Peter, 297
From Airline Reservations to Sonic the Hedgehog (Campbell-Kelly), 305

G

game theory, 45–48
Gates, Bill, 307
Grove, Andy, 256

H

Hamlet, Dick, 116, 307
Hemingway, Collins, 307
heuristics, 181–98
    abstraction, 186–98
    structured programming, 184
    successive refinement, 184–86
hierarchical teams, 34–38, 189
high-tech industry
    in developing countries, 106–7
    dot-com crisis and, 102–3
    in India, 103–4
    international perspectives on, 97–112
    in Israel, 104–6
    SARS and, 100
    September 11, 2001 attacks and, 99–100
    in the United States, 101–2
    women in, 107–10
    worldwide, 101–6
history
    early days of computing, 125–29
    of software development methods, 129–39
History of Modern Computing (Cerruzi), 305
Hoelzer, Helmut, 126–27
human aspects
    of software architecture, 193
    of software engineering, 187–88, 300–3
human-software interaction, debugging process and, 211–18
Humphrey, Watts, 25, 308
hypertext, e-commerce and, 257–58

I

IBM, 245–46
imitative programmers, 51
"Incredible Shrinking Pipeline, The" (Camp), 314
India, high-tech industry in, 103–4
"Inevitable Pain of Software Development, The" (Berry), 314
information hiding, 129, 134–35
Information Rules (Shapiro and Varian), 311
Innovation in Education (Intel), 106
interface prototypes, 133
international perspectives
    abstraction and, 191
    developing countries, 106–7
    Indian high-tech industry, 103–4
    Israeli high-tech industry, 104–6
    relevance of, 98–99
    on software engineering, 97–112
    on women in high-tech industries, 107–10
Internet
    communication and the, 261–62
    timeless nature of, 254, 259–62
Internet Time, 247
interviews, as data collection tool, 59–63
intuitive programmers, 51
Israel, high-tech industry in, 104–6
Iterative Model, 135–36

J

Jacobsen, Ivar, 138, 308
Johnson, Mark, 309

K

Kasher, Asa, 76
Kernighan, Brian, 309
Kerth, Norman, 309
Kessler, Robert, 313
knowledge management, 174–75
Knowledge Management Resource Center, 316
Kuhn, Thomas, 309

L

ladders of reflection, 164–70, 178
Lakoff, George, 309
learning organizations, 163, 171–76, 192
learning processes, 161–79
    abstraction and, 192
    reflective practice, 162–71
    team learning, 163
learning software organizations, 176
Lethbridge, Timothy, 299, 315
Linux, 246
Lister, Timothy, 306
Lovelace, Ada Augusta, 134

M

Making Truth (Brown), 305

Managing for the Future (Druker), 306
Manifesto for Agile Software Development, 119–20
Margolis, Jane, 310
marketing plan, 249
market share, 247
Mastering Virtual Teams (Snyder), 307
Mauchly, John, 127–28
Maybee, Joe, 116, 307
McConnell, Steve, 137, 310
mental models, 173
metaphors, 183, 195–96, 197, 256–57
Metaphors We Live By (Lakoff and Johnson), 309–10
Microsoft, 246
Miller, Roy, 304
modern systems theory, 21
Moore's Law, 245
Mythical Man-Month (Brooks), xxiii, 116, 245, 305

N

names, 209
NATO Conference (1968), xxii, 4, 130
Naur, Peter, 311
Naveda, J. Fernando, 306
"No Silver Bullet" (Brooks), 4, 14, 314

O

Object Modeling Technique (OMT), 138
objects, 138
observation, as data collection tool, 65–66
on-site customers, 19, 21
open-source software, 246
outsourcing
    Capability Maturity Model (CMM) and, 49–50
    virtual teams and, 38–39, 48–49
overtime, 228–29, 271

P

pair programming, 19, 165
Pair Programming Illuminated (Williams and Kessler), 313
Parnas, David, 129
Pascal, 4–5
Peopleware (DeMarco and Lister), 306
personal mastery, 173
Planning Game, 190, 235–36
Plauger, P.J., 309
PL/I, 4–5
power sources, 35–36
"Priorities for the Education and Training of Software Engineers" (Lethbridge), 315
Prisoner's Dilemma, 45–48, 190
process-oriented perspective, on software engineering, 117–19, 205

product-oriented perspective, on software engineering, 117–19
professions
    analogies to other, 121
    definition of, 76–77
program comprehension, 144–51
    abstraction and, 192
    actions for, 149–51
    factors affecting, 147–48
    mental models for, 149
    strategies for, 148
    theories of, 146–47
programmers
    gap between best and worst, 51–52
    intuitive vs. imitative, 51
programming
    See also eXtreme Programming
    bottom-up, 184
    languages, 134–35
    pair, 19, 165
    structured, 184
    styles, 207–11
    test-first, 21
project management
    See also software projects
    case studies, 271–75
    overtime and, 228–29
    schedule planning, 229–36
    slack time, 238–39
    using historical data, 229–30
Project Retrospective (Kerth), 309
prototypes, 135

Q

Quality Attribute Workshop (QAW), 195
questionnaires, as data collection tool, 63–65

R

Randell, Brian, 311
Rapid Application Development, 247
Rapid Development (McConnell), 310
Rational (company), 138
Rational Unified Process (RUP), 17, 138
Reading Computer Programs (Deimel and Naveda), 306
Recycling Principle, 277
refactoring, 21, 144–45, 154–57, 159, 210
Refactoring (Fowler), 307
Refactoring Home Page, 316
reflective practice, 21, 155, 162–71
Reflective Practitioner, The (Schön), 164, 311
requirements
    elicitation of, 236–38

management, 66–69
Ross, Douglas, 11, 130
Royce, Winston, 130–31
Rumbaugh, James, 138, 308
RUP (Rational Unified Process), 17, 138

S

SARS, 100
schedule delays, 9–10, 272–73
Schön, Donald, 121, 163–64, 311
Semi-Automatic Ground Environment (SAGE), 130
Senge, Peter, 171, 312
September 11, 2001, 99–100
Shapiro, Carl, 311
SHARE, 244
shared vision, 173–74
Shneiderman, Ben, 312
short releases, 21
simplicity, of code, 210
site, business, 248
Slack (DeMarco), 306
Snyder, Nancy Tennant, 307
software architecture
    agile software development and, 194–97
    human aspects of, 193
    vs. design, 193–94, 275–76
software business, 243–50
software characteristics, 203–7
    See also programming styles
software development
    See also software engineering
    case studies, 275–77
    complexity of, xxii
    customer requirements and, 55–69
    eXtreme Programming (XP), 18–22
    heuristics of, 181–98
        abstraction, 186–98
        structured programming, 184
        successive refinement, 184–86
    methods, 13–27
        abstraction, 129–30, 134–35, 188–89
        agile, 138–39
        choosing among, 22–24
        Evolutionary Development Model, 136
        history of, 129–39
        information hiding, 129, 134–35
        Iterative Model, 135–36
        life cycle, 130–32
        lightweight vs. heavyweight, 25
        objects, 138
        prototypes, 133, 135
        requiring, 24–25
        Spiral Model, 136–37
        Unified Process (UP), 138
        Unified Process (UP) of, 16–18
        Waterfall Life Cycle, 131–32, 135–36
    Spiral Model of, 15–16
    time-to-market, 247
software development teams, 29–53
    abstraction and, 189–90
    Bozo Effect and, 50–51
    Capability Maturity Model (CMM) and, 49–50
    Chief Programmer, 37
    democratic, 31–34
    game theory perspective on, 45–48
    hierarchical, 34–38
    perspectives on, 115
    Prisoner's Dilemma and, 45–48
    rewarding, 39–45
    student, 39–45
    Surgical Teams, 37–38
    types and structures of, 31–39
    virtual, 38–39, 48–49
software engineer
    conventional example of, 5–7
    progressive example of, 7–9
software engineering
    See also software development
    case studies, 269–81
    code of ethics, 73–96
    definitions of, 115–17
    history of, 125–39
    human aspects of, 187–88
    introduction of term, 4, 11, 130
    learning processes in, 161–79
    nonlinear nature of, 9
    perspectives on, 113–23
        agile vs. heavyweight, 119–20
        analogies to other professions, 121
        failure and success of software projects, 121–22
        international, 97–112
        product vs. process, 117–19
    as a reflective practice, 162–71
    search for best solution in, 4–5
Software Engineering Economics (Boehm), 304–5
software engineering education
    abstraction in, 197–98
    broadening, 295–96
    current state of, 298–300
    history of, 297–98
    relevance of, 297
    teaching human aspects in, 300–3
Software Engineering Ethics Research Institute, 315
Software Engineering Institute (SEI), xxiii, 49–50, 247, 297, 315

Software Engineering-Report on a Conference Sponsored by the NATP Science Committee (Naur and Randell), 311
software industry. *See* high-tech industry
software projects
    See also project management
    customer requirements for, 55–70
    estimating and tracking, 225–39
    life cycle of typical, 9–10
    poor management of, 226–36
    reasons for failure of, 4–5, 9–11
    slack time in, 238–39
    success or failure of, 121–22
Software Psychology (Shneiderman), 312
software requirements
    abstraction and, 190
    background on, 57–59
    changing, 55–57
    data collection tools for, 59–66
    management of, 66–69
spam, 262
Spiral Model, 15–16, 22–24, 136–37
Statement of Work (SOW), 249–50
status meetings, 234
stepwise refinement, 130
structured programming, 184
Structure of Scientific Revolutions (Kuhn), 309
student projects/presentations, 283–94
student teams, 39–45
successive refinement, 184–86
Surgical Teams, 37–38
SWEBOK (Software Engineering Body of Knowledge), xxiii
systems thinking, 172–73

T

team leaders, 35–36
team learning, 163, 174
teams
    Bozo Effect and, 50–51
    Chief Programmer, 37
    democratic, 31–34
    game theory perspective on, 45–48
    hierarchical, 34–38
    power sources in, 35–36
    rewarding, 39–45
    student, 39–45
    Surgical, 37–38
    virtual, 38–39, 48–49
Team Software Process (Humphrey), 308
Team Software Process (TSP), 24, 32, 36, 238
teamwork, 29–53
test-driven development, 195
Test Driven Development (Beck), 304
test-first programming, 21
top-down design, 184

U

Unified Development Process (Jacobsen et al.), 308
Unified Modeling Language (UML), 17, 18
Unified Process (UP), 16–18, 22–24, 138, 194
United States, high-tech industry in, 101–2
UNIVAC (Universal Arithmetic Calculator), 128
Unlocking the Clubhouse (Margolis and Fisher), 310

V

Varian, Hal, 311
velocity, 27
virtual teams, 38–39, 48–49
viruses, 262
VisiCalc, 246
vision, of product, 248
von Mayrhauser, Anneliese, 115
Von Neumann, 128

W

Wasserman, Anthony, 297
Waterfall Life Cycle, 15, 131–32, 135–36
Web sites, 315–16
"Why the Vasa Sank" (Fairley and Willshire), 314–15
Wilkes, Maurice, 128
Williams, Laurie, 313
Willshire, Mary Jane, 314
Winograd, Terry, 313
Wirth, Niklaus, 4–5
women, in high-tech industry, 107–10
Worldwide Institute of Software Architects, 121

X

XP. *See* eXtreme Programming

Y

Yourdan, Edward, 101, 313

Z

Zelkowitz, 115–16